AF510881

Proceedings

Tenth International Symposium on System Synthesis

Proceedings

Tenth International Symposium on System Synthesis

September 17–19, 1997
Antwerp, Belgium

Sponsored by

IEEE Computer Society
ACM SIGDA

In cooperation with

IFIP WG 10.5

IEEE COMPUTER SOCIETY

Los Alamitos, California

Washington • Brussels • Tokyo

IEEE Computer Society Order Number PR07949
IEEE Order Number 97TB100114
ACM Order Number 481970
ISBN 0-8186-7949-2
ISBN 0-8186-7951-4 (microfiche)
ISSN 1080-1820

Additional copies may be ordered from:

IEEE Computer Society
Customer Service Center
10662 Los Vaqueros Circle
P.O. Box 3014
Los Alamitos, CA 90720-1314
Tel: + 1-714-821-8380
Fax: + 1-714-821-4641
E-mail: cs.books@computer.org

IEEE Service Center
445 Hoes Lane
P.O. Box 1331
Piscataway, NJ 08855-1331
Tel: + 1-908-981-1393
Fax: + 1-908-981-9667
mis.custserv@computer.org

IEEE Computer Society
13, Avenue de l'Aquilon
B-1200 Brussels
BELGIUM
Tel: + 32-2-770-2198
Fax: + 32-2-770-8505
euro.ofc@computer.org

IEEE Computer Society
Ooshima Building
2-19-1 Minami-Aoyama
Minato-ku, Tokyo 107
JAPAN
Tel: + 81-3-3408-3118
Fax: + 81-3-3408-3553
tokyo.ofc@computer.org

ACM
1515 Broadway, 17th Floor, New York, NY 10036-5701
Tel: 1 212 626-0500, Tel: 1-800-342-6626, Fax: 1-212-944-1318, Email: orders@acm.org

Editorial production by Penny Storms

Cover art production by Alex Torres

Printed in the United States of America by The Printing House

IEEE COMPUTER SOCIETY

IEEE

Table of Contents

Message from the Program Chair

To keep up with growing system complexities and shrinking time-to-market constraints, designers are shifting focus to ever higher levels of system design abstraction, using rapidly-maturing exploration and synthesis tools to help automate and support the lower-level details. ISSS has assisted this shift by serving as the premier international forum for the leading, or more precisely, the "highest" edge of synthesis research. It has always tried to provide a good balance between the academic (theoretical research) and industry (relevant results) arenas. The meeting began as the High-Level Synthesis Workshop in the early-1980's, and evolved into a symposium in 1994. After over a decade of research, the first commercial high-level (or behavioral) synthesis tools appeared in the mid-1990's.

With the growing maturity and acceptance of high-level synthesis, and the unified view of both hardware and software that such tools enable, the symposium expanded and even moved to higher-levels of abstraction and new technologies and design issues. Presently it covers: system-level and architectural synthesis, hardware/software codesign, programmable (multi-)processor-based design, system-design experience and methodologies, embedded and real-time system software, and synthesis for low power and testability and verifiability. The symposium also took on a new name in 1995 to better reflect its new focus on "system" issues: the International Symposium on System Synthesis.

Now in its tenth anniversary, ISSS serves as a forum for mature research results as well as for promising new ideas. This year we had 60 submissions from 17 countries, which makes ISSS'97 a truly international event. The program committee selected 18 excellent contributions from those submissions following a rigorous review process. Each paper was evaluated by 4 to 6 independent reviewers, and results were accumulated through a fully-electronic, mostly web-page-based review process, resulting in an initial classification of papers into "Clear Accept," "Middle," and "Clear Reject" categories. Several further rounds of electronic consultation on the middle papers were carried out by the program committee members until full consensus was reached on the final selections. Due to the successful introduction of a web-page-based submission and the emphasis on electronic handling, only 1 paper was still submitted on plain paper.

ISSS'97 also features 6 invited tutorial talks by leading industrial (Dr. Frank Gielen, Alcatel-Telecom; Dr. Roelof Salters, Philips; Dr. Richard Taylor, HP Labs) and academic experts (Prof. Hugo De Man, Prof. Gerry Musgrave, Prof. Alex Nicolau) on very timely and interesting topics related but not overlapping with the other ISSS activities. In addition, there are 2 panel sessions on topics of active interest for the future of our community.

Finally, I would like to thank the many committee members who contributed to the success of this symposium. Foremost, I would like to thank our Publicity Chair, Diederik Verkest, for his major contribution in setting up the web page, especially for the automation of the review process and electronic paper submission handling. The result was unanimously appreciated by the "users" and it is certainly reusable for future occasions. I would also like to give special thanks to our Finance Chair, Annemie Stas, and our Publications Chair, Edwin Sha, for their significant effort in setting up these issues in the organization. Our General Chair, Frank Vahid, and the previous Publicity Chair, Sanjiv Narayan, provided us with a wealth of information about the past year's event, a very useful and true inheritance. The help of Ivo Bolsens and Hugo De Man in selecting and attracting the excellent invited speaker program is also appreciated. All the steering committee members were of great assistance in planning the symposium. Last, but not least, I would like to thank the program committee for their tremendous efforts in thoroughly reviewing the papers and in updating their review input during the many iterations, which resulted in a smooth convergence.

Further ISSS information can be found at: http://www.imec.be/isss97/, and by joining an email list by sending email to: isss-requests@ics.uci.edu.

Enjoy this 10th anniversary event and its beautiful location: Antwerp!

Francky Catthoor
Program Chair

ISSS'97 Organizing Committee

General Chair
Frank Vahid, *UC Riverside*

Program Chair
Francky Catthoor, *IMEC*

Publicity Chair
Diederik Verkest, *IMEC*

Finance Chair
Annemie Stas, *IMEC*

Publications Chair
Edwin Sha, *University of Notre Dame*

Panels Chair
Rudy Lauwereins, *K.U. Leuven*

Past Chair
Ahmed Jerraya, *TIMA-CNRS, Grenoble*

Technical Program Committee

Marleen Adé, *K.U. Leuven, Belgium*
Gaetano Borriello, *University of Washington, USA*
Raul Camposano, *Synopsys, USA*
Nikil Dutt, *University of California, Irvine, USA*
Rolf Ernst, *Technical University Braunschweig, Germany*
Daniel Gajski, *University of California, Irvine, USA*
Cathy Gebotys, *University of Waterloo, Canada*
Yu-Chin Hsu, *Avant!, USA*
Kayhan Kucukcakar, *Motorola, USA*
Fadi Kurdahi, *University of California, Irvine, USA*
Steve Y.L. Lin, *Tsing Hua University, Taiwan*
Paul Lippens, *Philips, The Netherlands*
Jan Madsen, *Technical University Denmark, Denmark*
Lev Markov, *Mentor Graphics, USA*
Peter Marwedel, *University of Dortmund, Germany*
Vijay Nagasamy, *LSI Logic, USA*
Yukihiro Nakamura, *NTT, Japan*
Sanjiv Narayan, *Ambit Design Systems, USA*
Kevin O'Brien, *Leda, France*
Pierre Paulin, *SGS-Thomson, France*
Wolfgang Rosenstiel, *University of Tubingen, FZI, Germany*
Edwin Sha, *University of Notre Dame, USA*
Leon Stok, *IBM, USA*
Donald Thomas, *CMU, USA*
Kazutoshi Wakabayashi, *NEC, USA*
Robert Walker, *RPI, USA*
Wayne Wolf, *Princeton University, USA*

Reviewers

Marleen Ade
Gaurav Aggarwal
Hiroki Akaboshi
Steven Bashford
Peter Björn-Jörgensen
Christian Blumenroehr
Gaetano Borriello
Oliver Bringmann
Randy Bryant
Tim Buchholz
Raul Camposano
Antenor Carvalho
Francky Catthoor
Chantana Chantrapornchai
Chih-Tung Chen
Vincent En-Shou Chang
Wei-Kai Cheng
Pai Chou
Johan Cockx
Marco Cornero
Koen Danckaert
Ali Dasdan
Luc Decoster
Jean-Philipe Diguet
Eddy De Greef
Rainer Doemer
Nikil Dutt
Dirk Eisenbiegler
Rolf Ernst
Nong Fan
Michel Favre
Dan Gajski
Cathy Gebotys
Frank Ghenassia
Jie Gong
Peter Gabi Grun
Ashok Halambi
Andri Hergenhan
Dirk Herrmann
Ken Hines
Yu-Chin Hsu
Ahmed Jerraya
Hsiao-Ping Juan
Jon Kleinsmith
Peter Voigt Knudsen
Gernot Koch
David Kolson
Kayhan Kucukcakar
Fadi Kurdahi
Rudy Lauwereins
Mike Lee
Eric Lin
Youn-Long Lin
Paul Lippens
Ta-Yung Liu
Jan Madsen
Gilberto Marchioro
Lev Markov
Peter Marwedel
Hans-Georg Martin
Akihiro Matsuura
Vasily Moshnyaga
Kohichi Nagami
Hidehisa Nagano
Vijay Nagasamy
Akira Nagoya
Yukihiro Nakamura
Sanjiv Narayan
Francisco Nascimento
Ryo Nomura
Steven Novack
Kevin O'Brien
Ross Ortega
Preeti Panda
Pierre Paulin
Wim Philipsen
Salil Raje
Wolfgang Rosenstiel
Miguel Santana
Hiroshi Sawada
Edwin Sha
Tsunemiti Shiozawa
Hartej Singh
Peter Slock
Leon Stok
Marino Strik
Takayuki Suyama
Katsuharu Suzuki
Weiyu Tang
Filip Thoen
Don Thomas
Thomas Tkacik
Kevin Tsai
Paul Van Besow
Jan Van Der Steen
Maurizio Vitale
Kazutoshi Wakabayashi
Robert Walker
Thomas Wang
Karlheinz Weiss
Wayne Wolf
Tsung-Yi Wu
Sven Wuytack
Min Xu
Shigeru Yamashita
Zhihong Yu
Yang Zhang
Jianwen Zhu

Session 1

Formal Specification and Validation

Quick Conservative Causality Analysis

Ellen M. Sentovich
Cadence Berkeley Laboratories
1919 Addison Street Suite 303-304
Berkeley, CA 94704-1144

Abstract

The causality problem is that of determining if a combinational circuit with cycles has acceptable behavior: that for all inputs the outputs are well-defined and stable. While the problem manifests itself at the **circuit** *level, it usually originates at the* **system** *level. It may arise when a system is designed as a collection of modules: when composed, cycles are discovered in the ensemble. One must analyze these cycles to correct possible errors or to capture the correct behavior appropriately for further synthesis. Previously published algorithms use iterated ternary logic simulation. This is correct and robust, but expensive and in many cases overkill. In this work, a more efficient but conservative algorithm is proposed based on applying standard logic synthesis techniques of increasing power. We present initial results to demonstrate the practicality of this approach.*

1. Introduction

1.1. The causality problem

The *causality* problem here is that of determining if a combinational circuit with cycles has *acceptable* behavior. Acceptable circuit behavior is often defined as behavior that has an input/output-equivalent acyclic circuit implementation. That is, for each input of the cyclic combinational circuit, the outputs are well-defined and stable. Any algorithm for causality analysis relies on a particular definition of acceptable behavior, which specifies equivalence and stability conditions. This behavior in turn relies on an underlying delay model. At a higher level, causality can be viewed as the property that a module has well-defined, stable behavior at the outputs for each possible input.

The causality problem may arise in several contexts:

- An *implicit* specification of a single circuit or state machine may not have a well-defined causal implementation (i.e., with an acyclic combinational part).
- A set of causal circuits composed together may exhibit non-causal behavior.
- In system specification, where the design is composed of individual modules, combinational loops between modules may be accidentally or deliberately (e.g., shared resources) designed in.

In many cases, the design has been well-conceived, and the cycles are in fact false: they are static cycles that are never active dynamically.

With ESTEREL[2] specifications, causality problems can arise in two ways. First, due to the powerful language constructs and implicit nature of state-machine specification, an individual module may exhibit causality problems. Second, during module composition, individually causal modules may produce a system with causality problems. A number of examples of the first type, some causal and some non-causal, are given in [1]. An example of the second type, which is a real bus arbiter that is causal but cyclic, is given in [4]. The specification has a circuit implementation that is statically cyclic but dynamically acyclic. Causality analysis is used in the compiler now to determine which systems are truly causal despite their structural cycles.

Another application is in behavioral specifications, where resources may be shared in such a way as to create a static cycle in the specification. An example is given in [6]: two operators ADD and SHIFT may be performed in either order depending on the value of a select variable. Hence there are paths from ADD to SHIFT and from SHIFT to ADD, but they are never active simultaneously.

When combinational cycles are detected, causality analysis is usually carried out before synthesis and optimization proceeds. If the circuit is correct (causal), it can be re-implemented as an acyclic circuit. If incorrect, one must find the undesirable behavior and correct it. One can certainly generate cyclic hardware and software implementations for cyclic specifications immediately, skipping causality analysis. However, without this analysis, there is no guarantee of correct behavior, and without generation of the equivalent acyclic implementation, there are currently no methods for

1080-1820/97 $10.00 © 1997 IEEE

synthesis and optimization.

1.2. Previous work

In [6], a definition of combinational circuits[1] was given with an algorithm for determining if a cyclic circuit is in fact combinational. The algorithm is based on symbolic simulation using the 3-valued Scott Boolean domain: $\{0, 1, \perp\}$. It consists of breaking all cycles in the circuit, assigning broken wires the undefined value $\perp$, and iteratively simulating, assigning newly-determined simulation values to the broken feedback wires at each iteration. The algorithm is monotonic and guaranteed to converge. This paper also provides good motivation and references for the causality problem (e.g., [11] which solves the problem of cycles in the context of resource sharing by restricting the sharings considered).

In [8], the algorithm was extended to sequential circuits, and an efficient implementation was outlined. In [9], a proof of correctness of the algorithm under the up-bounded inertial delay model was given.

In parallel, the notion of *constructive causality* [1] was being developed and imposed as a requirement for correct ESTEREL specifications. In the ESTEREL language, one specifies a synchronous, reactive, control system as a set of interacting modules. The language is based on communication of signals, and provides constructs for concurrency and pre-emption. The ESTEREL compiler translates the set of modules into an equivalent Boolean circuit, which is another implicit representation of the underlying state machine.

The implicit specification method and rich signal constructs imply that non-causal behavior can be given. The compiler performs a static causality check on the input program translated to a dependency graph. This approximation to the definition of causality held at the time turned out to be too weak. As a result, constructive causality was born. Its complete semantics at the behavioral, operational, and circuit levels, is given in [1]. The basic idea is that a design must be *reactive* (have at least one response per input), *deterministic* (at most one response), and *constructive* (the solution is derived by fact propagation rather than self-justification). The key result is that *a constructively causal design has a stable, well-defined circuit equivalent.* Therefore, the complete causality analysis program in ESTEREL, `sccausal` [5], is based on the results in [6, 8, 9, 1] and is applied to the Boolean circuit representation of ESTEREL programs.

[1] A *combinational circuit* hereafter is a circuit with no delay elements that has "acceptable" behavior, as described in the beginning of this introduction and formally defined in [9].

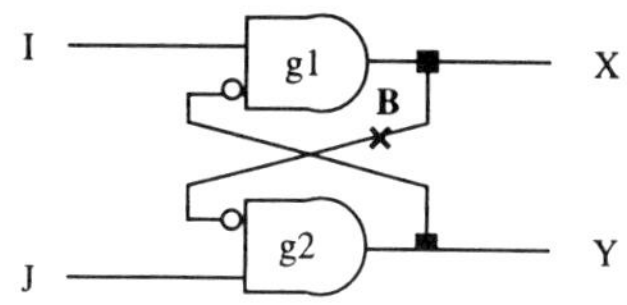

Figure 1. Simple non-causal circuit

1.3. Efficient causality analysis

We present here an algorithm that is built on the same assumptions (delay model, ternary simulation) but that uses standard logic synthesis and optimization. It is more efficient since it does not require a three-valued model, it does not perform iteration, and it uses a variety of efficient algorithms applied successively to the circuit rather than full symbolic simulation at the outset. It is also conservative in that it may declare a causal circuit non-causal. We focus on the analysis of cyclic combinational circuits to determine an acyclic equivalent. The extension to sequential circuits is not trivial, but has been done in [8]; that method can be easily modified to work with the causality analysis method proposed here. Of course, combinational causality analysis can be applied to sequential circuits directly, but the analysis will only be carried out statically: as though all states are reachable. We present results on real combinational and sequential circuits generated from ESTEREL.

2. Example

A simple example illustrates the notion of causality, causality analysis, previous algorithms, and the contribution of this work.

Consider the circuit shown in Figure 1. It is electrically non-causal: when I and J are both 1, $X = \bar{Y}$ and $Y = \bar{X}$. While stable solutions exist ($X = 1$, $Y = 0$; $X = 0$, $Y = 1$), there is no guarantee that the wires will stabilize given the delays, and there is no way of knowing which of the two stable solutions will be reached. The ternary simulation algorithm would break the cycle at, for example, point B, creating a new input and output, B_{in} and B_{out}. Assigning B_{in} to $\perp$ (unknown), and allowing the inputs to assume all values, we obtain $B_{out} = I \cdot \bar{Y} = I \cdot (\bar{J} + B_{in}) = I\bar{J} + I\perp$. Iterating, $B_{out} = I \cdot (\bar{J} + (I \cdot \bar{J} + I \cdot \perp)) = I\bar{J} + I\perp$. There is no change, so the iteration stops and the circuit is non-causal since B_{out} cannot be assigned a stable value for each value of the input.

Now consider a modification of this circuit, where the $\{1, 1\}$ input to the cross-coupled gates is prohibited by the environment. This is modeled by the circuit in Figure 2. Again breaking the cycle at B, $B_{out} = I'\bar{Y} = I' \cdot (\bar{J}' + B_{in}) = I\bar{J} \cdot (I + \bar{J} + B_{in}) = I\bar{J} + I\bar{J}\perp = I\bar{J}$ and stability is

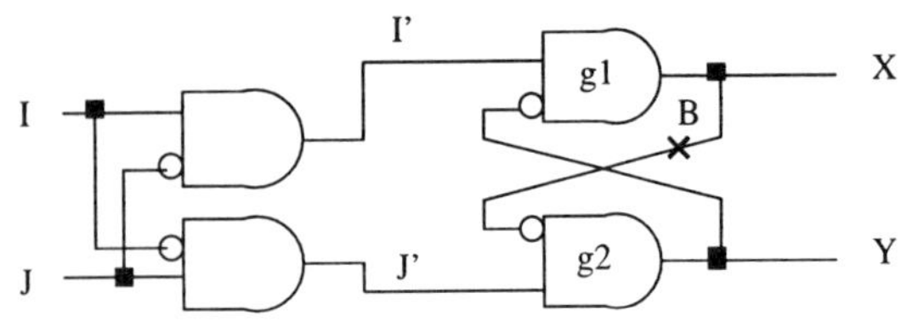

Figure 2. Restricted inputs creates a causal version

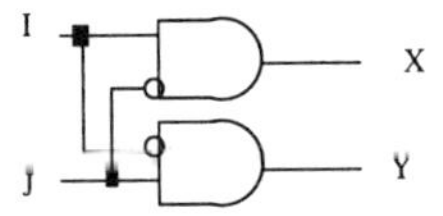

Figure 3. Acyclic version

reached. B_{in} is now implemented by $I\bar{J}$. The final acyclic version of the circuit is shown in Figure 3.

Now suppose we take the original circuits, cut the B arcs, and ask the question "is B_{out} independent of B_{in}?" For the non-causal circuit, $B_{out} = I\bar{J} + IB_{in}$, $B_{out\,B_{in}} \overline{\oplus} B_{out\,\overline{B_{in}}} = \bar{I} + \bar{J} \neq 1$ and the answer is no. For the causal circuit, $B_{out} = I\bar{J}$ and the answer is yes. This simple analysis is equivalent to a single iteration of the ternary symbolic simulation, but note that there are many ways of determining the independence of B_{out} from B_{in} using logic synthesis techniques. For example, in the simplest case, constant propagation in the circuit may reveal independence. Efficient ATPG techniques can be used to determine independence, as can logic simplification. In the limit, the BDD for B_{out} can be built and tested for independence: $B_{out\,B_{in}} \equiv B_{out\,\overline{B_{in}}}$? Or even more simply, does B_{in} appear as a variable in $BDD(B_{out})$? The independence check is of course computationally equivalent to exhaustive simulation of an output with respect to an input, but by viewing the problem this way, and noting that in many cases full analysis is not required, a better algorithm can be devised. Simple logic synthesis algorithms are not performed a priori in the ternary simulation analysis method because an alteration of the logic may destroy the causality properties of the circuit.

Such simple arc-breaking followed by dependency analysis is not by itself a correct algorithm for causality determination, as the next example illustrates.

Example 2.1 *Consider the following simple non-causal circuit.*

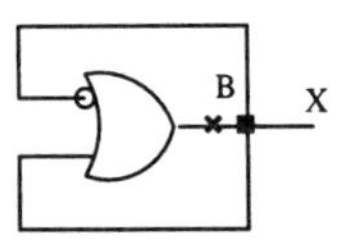

This circuit is reactive and deterministic, and hence logically correct. However, it is not constructively causal since it stabilizes only due to feedback. See [1] for the precise definition of constructive causality and the reasons behind this choice.

Breaking the two cycles by cutting before the fanout point results in $B_{out} = B_{in} + \overline{B_{in}} = 1$, which is not the correct result. Breaking the two cycles using the two feedback arcs does retain the dependency needed to demonstrate that this circuit is non-causal. The difference, of course, is precisely what one obtains using ternary versus binary simulation.

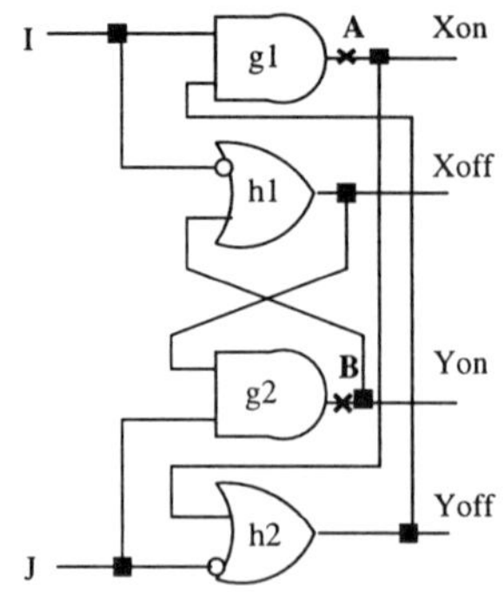

Figure 4. Dual-rail encoding of IJ-example

3. Method

Several observations about causality analysis and ESTEREL-generated circuits lead to our new algorithm, as described in this section.

Observation 3.1 *Ternary simulation is equivalent to dual-rail encoding and binary simulation.*

This is formally proven in [3]. This observation implies that in the complete causality analysis algorithm, one can replace a ternary simulation iteration by a transformation of the circuit to the dual-rail encoded version followed by a binary simulation. The dual-rail encoded version of the IJ-example circuit is shown in Figure 4. Note that each internal signal is encoded by a pair of signals, and that all the gates in the circuit are positive unate (no negations) with respect to the internal signals. In practice, it is most efficient to first break the cycles and dual-rail encode only the signals corresponding to the broken cycles, and all the signals in their transitive fanout cones.

Example 3.1 *The dual-rail encoding of the OR-gate with feedback is given by*

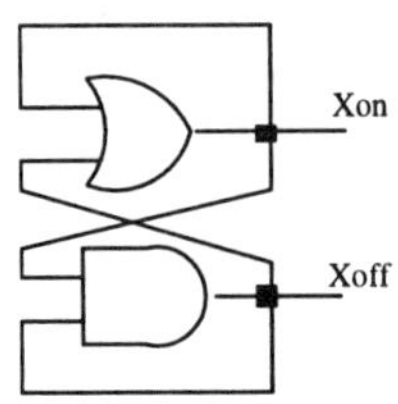

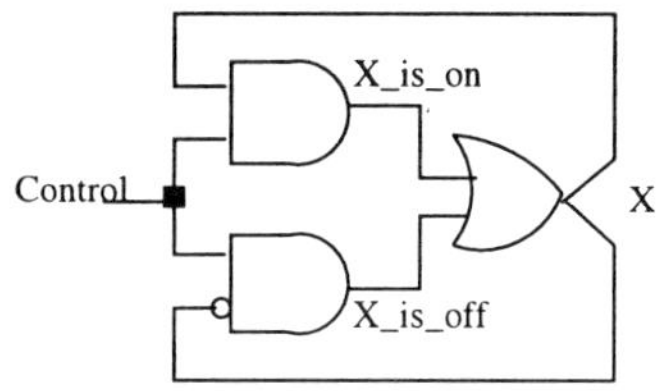

Figure 5. ESTEREL version of OR-gate example

The reader can verify that cycle breaking and dependency analysis will now yield the correct result: Xon_{out} *and* $Xoff_{out}$ *will depend on* Xon_{in} *and* $Xoff_{in}$ *respectively, regardless of where the feedback is broken.*

Observation 3.2 *Often in practice iteration is not necessary, especially if one breaks the feedback judiciously.*

This observation indicates that a non-iterating, conservative computation will often suffice to correctly compute causality. In [6], it was demonstrated that convergence is reached in $\leq k$ iterations, where k is the number of feedback arcs broken. It follows that if only one arc is broken, the non-iterating algorithm produces the correct result.

Thus, it is very important to choose the feedback set carefully. If simulation is iterated, a poor choice still leads to a correct result, but after many iterations. (For example, consider the case of breaking all the internal wires of a circuit.) If simulation is not iterated, fewer feedback arcs provides increased reliability of results as less information flows across the feedback boundary.

Observation 3.3 ESTEREL-*generated circuits are partially dual-rail encoded.*

The OR-gate example treated in the previous section might be specified in ESTEREL as follows:

```
module OR:

signal X in
present X then emit X else emit X end
end signal

end module
```

The direct translation to a circuit as described in [1] leads to the circuit shown in Figure 5.[2] While Xon and Xoff are not explicitly modeled by wires as they are for full dual-rail, the positive and negative tests for X being ON and OFF are explicitly modeled by wires X_is_on and X_is_off as shown in the figure. The present test of a signal in ESTEREL *always* generates two gates in this manner and hence two fanout arcs. Note also that fanout points are not modeled: gates have multiple fanout arcs. If the signal is on a cycle and its fanout does not reconverge, effectively both arcs must be broken to break all feedback (which is equivalent to dual-rail). If there is reconvergence, it is possible that only one arc is broken (where two must be broken in the dual-rail version). This may still return a correct causality result depending on the function at reconvergence, *but not always*. The circuits in Figures 1 and 2, for example, are correctly analyzed with one broken feedback arc. Still, we can take advantage of this pseudo-dual-rail property of ESTEREL circuits to build and test a quick prototype causality checker. In summary:

[2] This is only part of the circuit generated. There is additional circuitry implementing initialization and the termination codes.

- Observation 3.1 implies that an iteration of the ternary simulation-based causality checking algorithm can be performed by first dual-rail encoding the circuit and then performing binary simulation. Furthermore, any binary-based transformations (e.g., standard logic optimization) can be performed on the dual-rail version without destroying causality properties.
- Observation 3.2 implies that in many cases, and certainly if only one arc is broken, a single simulation iteration suffices to determine causality. If multiple iterations are required for a causal circuit, at the completion of the first iteration the information is simply inconclusive: it will appear as though the values of the feedback arcs are unknown, when in fact further iteration would determine these values. Hence, single-iteration simulation is usually correct by observation 3.2, and at least conservative (a non-causal conclusion may be made for a causal circuit).
- Observation 3.3 implies that in many cases, one need not fully dual-rail encode a circuit to use the binary simulation method for causality determination: the ESTEREL-generated circuit already duplicates enough signals that arc-breaking and binary simulation suffices to determine causality. The result in this case is not conservative as a non-causal circuit may be deemed causal. However, one can use this notion to implement a prototype causality checker to compare the performance of a binary-simulation-based algorithm to that of the full ternary one.

These are the main ideas behind our method. The algorithm is described more thoroughly in the next section.

4. Algorithm

Observations from the previous section indicate that a conservative alternative algorithm to iterated ternary simulation would consist of:

- breaking a minimal number of feedback wires
- transforming the circuit into the dual-rail encoded version by dual-rail-duplicating gates (adding the DeMorgan-equivalent gate) in the transitive fanout of each broken-arc input
- performing dependency analysis (binary simulation).

Observation 3.3 only indicates that one may omit the dual-rail encoding step and still obtain the correct result in some cases. (In all of our cases, we obtain the correct result, both for causal and non-causal circuits.) We use this observation to quickly build a prototype of our algorithm to test it against the current version, though in practice this is of less use since it is not conservative, and thus conservative or full causality will have to be run subsequently anyway (except in the cases where it determines a circuit to be non-causal, and produces a meaningful error trace). The prototype algorithm simply

1. breaks cycles
2. performs a dependency computation
3. creates an acyclic equivalent version if possible, or produces an input pattern exhibiting the causality error.

4.1. Breaking combinational cycles

As noted in Section 3, it is desirable to break as few arcs as possible. We implemented two algorithms for cycle-breaking. The first is guaranteed to break the minimum number of arcs, and was published in [10]. It is very fast even on large graphs. The second is the very simple but fast depth-first search algorithm of Tarjan. In our experiments, as we had quite small examples or large examples with small strongly connected components, we always used the more robust exact algorithm.

4.2. Dependency computation

The second step in the algorithm is to determine whether or not the broken feedback outputs logically depend on the broken feedback inputs; if not, they can be expressed independently, and the broken arcs reconnected to create an acyclic equivalent circuit.

The dependency computation interleaves logic optimization and dependency check. The simple dependency check processes nodes in topological order searching for a path from each B_{in} to each B_{out} and using this information to compute on-the-fly whether cycles would appear in the circuit were the arcs to be reconnected. If not, the check is successful. If only a single arc is broken, this computation simplifies to a check for a path from B_{in} to B_{out}. The complete dependency computation proceeds as follows:

- simple constant propagation; simple dependency check
- simple logic optimization, collapsing small nodes together and simplifying node functions; simple dependency check
- introduce external don't care conditions from the ESTEREL specification and repeat simple logic optimization; simple dependency check
- build BDDs for all B_{out} variables; check for BDD dependencies of each B_{out} on its corresponding B_{in}, determine the existence of cycles.

The last is the most robust, and returns an exact answer as to whether an output depends on an input. It is not too expensive in practice, as the strongly connected components in the original circuit graph are small (and hence the logic cones between temporary inputs and outputs), and the circuit has already been optimized before building the BDDs.

In general, it is not sufficient to check simply that each B_{out} is independent of its corresponding B_{in}. With more than one feedback arc, interdependencies could still lead to a cyclic circuit on reconnection. For this, it suffices to build a dependency graph using the dependency information computed above (from simple to robust) and perform acyclicity checks on this graph. The on-the-fly cycle check mentioned above is an efficient implementation of this. In practice, we never observed interdependencies leading to additional cycles.

4.3. Creating the acyclic circuit or producing an error trace

If the simple dependency check is successful, an acyclic version of the circuit can be created by simply reconnecting the broken feedback arcs. A negative simple dependency check result implies there is no path from the input to the output of interest (or no cycle-forming set of paths), so simple reconnection will result in an acyclic circuit. (This will only happen after some logic alteration, such as constant propagation. Before this, it is known that there is at least one topological path between the two or there would not have been a cycle.)

If simple dependency fails but BDD dependency succeeds, the acyclic circuit must be created using logic functions derived from the BDD and then reconnecting the feedback arcs. That is, the function for each feedback arc output is completely re-implemented based on its BDD. This step can imply a significant logic increase depending on the functions: the ADD/SHIFT example mentioned in the introduction would require duplication of these operators. In practice, this step is not too expensive as the BDDs are limited to only the necessary scope in the circuit. Furthermore, our examples were limited to circuits described in ESTEREL, which tend to be control-based.

If the circuit is dual-rail encoded before the dependency check, a circuit equivalent to the original is obtained by merging the dual rail (feedback) inputs and removing the negative dual rail outputs before reconnecting the feedback arcs. This is followed by simple logic optimization and/or explicit merging of the duplicated internal signals to recover the area overhead incurred in dual-rail encoding.

If the circuit is determined to be non-causal, an error condition is produced simply by: $Input\_error = B_{out\,B_{in}} \oplus B_{out\,\overline{B_{in}}}$. For sequential circuits, an additional constraint is imposed that the *Input_error* contain the initial state, to ensure that it is a valid condition.[3]

4.4. Optimizing cyclic circuits

We have cast the algorithm as one for causality analysis, but it can be used as a cyclic circuit optimizer as well. Suppose one has a cyclic implementation, and one would like to retain this form since it is more efficient. There are currently no logic optimization programs that can optimize cyclic circuits. In addition, for the same reason that ternary simulation must be used to perform causality analysis, acyclic subcircuits of a cyclic circuit cannot be abstracted and optimized with standard techniques while maintaining the causality properties. The simple algorithm proposed here is a cyclic circuit optimizer: cycles are dual-rail encoded, standard logic optimization is applied, dual-rail signals and gates are merged to recuperate the overhead, and a final logic optimization is applied.

5. Implementation and results

The program `cheap_cause` has been implemented inside the SIS logic synthesis program [7]. A dependency graph is created for the cyclic circuit and used to compute the set of feedback arcs. The corresponding wires are broken in the network and the dependency analysis performed. Dual-rail encoding was not implemented for this prototype implementation; in fact, with the simple reliance on the partial dual-rail encoding obtained with the ESTEREL translation, the correct causality result was obtained in all cases. Furthermore, the advantage of this method in terms of final logic area will not be affected by the overhead incurred by dual-rail encoding: duplicated gates and signals can later be merged.

[3]It may be valid in another state, but we do not yet do the analysis to determine the valid reachable states. In all experiments, an error condition was found at the initial state.

5.1. Comparison with `sccausal`

For comparison, we describe the `sccausal` algorithm [5] and note some of its properties. First, three-valued functions (TVFs) for each node computed using two variables per node, and using a BDD representation. This is expensive since the number of variables is doubled; it has a dramatic effect on the efficiency of the BDD computations. Next, a weak topological ordering (WTO) is computed for the nodes: rather than compute feedback arcs directly, an ordering is computed for node processing during simulation. This ordering is not a static one-pass through all nodes, but rather contains cycles within that are iterated to convergence. The algorithm iterates to convergence, correctly handles full ESTEREL (the algorithm described here handles only pure ESTEREL), and handles sequential circuits. Finally, if the circuit is causal, an acyclic version is built based on the BDDs. This is another source of inefficiency, since the structure of the initial implementation is lost completely.

5.2. Experiments

A number of experiments have been run on the prototype version of `cheap_cause`, and comparisons made with `sccausal`. However, we note the following points. First, neither `sccausal` nor `cheap_cause` are fully optimized: `sccausal` has been released in a beta version, and `cheap_cause` is still a prototype. Second, though in some cases `cheap_cause` may return the correct result immediately (in all cases, in our examples), even with full dual-rail encoding it is conservative since it performs no iteration. Therefore, its utility is intended more as a preprocessor to `sccausal`, and as an optimization scheme for cyclic circuits, than as a causality analysis program in its own right.

We have run our program on all the anomalous causality programs `P1–P13` described in [1]. In all cases, `cheap_cause` returned the correct result very quickly. Comparison of computation times with `sccausal` is not meaningful since the circuits are so small.

The program was applied to an industrial example for in-vehicle communication [12] called **prosa**. The original version is causal but has cycles. `sccausal` makes this determination and computes the acyclic circuit in 720 seconds on a sun4. `cheap_cause` was able to correctly determine this with only one broken arc (the original graph of 1989 vertices and 3463 edges is reduced to a single strongly connected component with 82 vertices and 118 edges), and simple constant propagation, in 19.4 seconds. The circuit produced by `sccausal` has 8346(6842) literals in sum-of-products (factored) form, and 64 latches; after optimization in SIS, it has 1099(839) literals and 44 latches. The circuit produced by `cheap_cause` has 1268(1268) literals and 73 latches; after optimization, 838(761) literals and 71 latches.

On a second, non-causal version of **prosa**, `sccausal` completes in 987 seconds, `cheap_cause` in 61 seconds. The error traces cannot be compared since `sccausal` returns values on internal **sc** format wires and corresponding lines in ESTEREL code (useful for source-code debugging), and `cheap_cause` returns values on the circuit I/O signals (useful in the ESTEREL simulators).

The second example, **mejia**, is an industrial control design which is causal but contains cycles. `sccausal` performs the analysis and writes the result in 44 seconds. The circuit has 2910(2454) literals and 28 latches, reduced to 637(484) literals and 21 latches by SIS. `cheap_cause` breaks one arc and performs the analysis in 25 seconds, with a resulting circuit of 878(878) literals and 43 latches, reduced to 543(439) literals and 43 latches by SIS.

It is interesting to note the difference in the number of registers. `sccausal` is able to reduce this number on the fly during the reachable states computation. In both cases, the initial circuit is considerably smaller for `cheap_cause`, which could make a decidable difference for further optimization of large circuits.

These are of course good scenarios on the type of designs that `cheap_cause` can handle. We mention these results simply as indication that the proposed algorithms have practical use; more experimentation is warranted.

6. Conclusions and future work

We have presented an algorithm for efficient causality checking. While it is conservative, it has been demonstrated on a suite of examples to return correct results. It is based on the observation that ternary simulation is equivalent to dual-rail encoding of feedback signals followed by binary simulation. The latter is implemented efficiently by applying a series of logic optimization techniques of increasing power. The results provide both a method for quick causality analysis, and a method for optimization of cyclic circuits. Furthermore, the final acyclic implementation may be produced directly from the logic synthesis tools rather than from BDDs, so it retains the structure of the initial implementation as much as possible.

There are several areas for future work. First, full cheap causality needs to be implemented with complete dual-rail encoding of the circuits, rather than relying on the structure of ESTEREL-generated circuits and the arc-breaking algorithm to maintain some integrity of the causality analysis. Second, a modified algorithm must be developed for sequential circuits. This algorithm will simply be an iteration of the causality analysis presented here, and a reachable states computation (both done on the arc-broken acyclic circuit) similar to the algorithm published in [8]. Third, more examples for causality analysis must be obtained and analyzed, so that the gain of this new method over the complete method can be assessed on practical designs. Finally, the techniques should be applied to cyclic circuits as an a priori optimization method, before full-causality is carried out. This should improve the performance of `sccausal` significantly.

Acknowledgements

Helpful discussions with Gérard Berry are gratefully acknowledged. This work was supported in part by the National Science Foundation under grant INT-9505943.

References

[1] G. Berry. *The Constructive Semantics of Pure Esterel.* 1996. To Appear, available now at ftp: //www.inria.fr/meije/esterel/papers/constructiveness.ps.gz.

[2] G. Berry and G. Gonthier. The Esterel Synchronous Programming Language: Design Semantics, Implementation. *Science of Computer Programming*, 19(2):87–152, 1992.

[3] G. Berry and E. Sentovich. On constructive causality, 1997. Work in progress.

[4] R. DeSimone. Note: A small hardware bus arbiter specification leading naturally to correct cyclic description, Mar. 1996. Technical note.

[5] A. Girault, T. Shiple, and H. Toma. The sc-causal compiler, 1997. Documentation provided with the ESTEREL compiler.

[6] S. Malik. Analysis of Cyclic Combinational Circuits. In *Proceedings of the IEEE International Conference on Computer-Aided Design*, pages 618–625, Nov. 1993.

[7] E. Sentovich, K. Singh, C. Moon, H. Savoj, R. Brayton, and A. Sangiovanni-Vincentelli. Sequential Circuit Design Using Synthesis and Optimization. In *Proc of the ICCD*, pages 328–333, Oct. 1992.

[8] T. Shiple, G. Berry, and H. Touati. Constructive Analysis of Cyclic Circuits. In *Proceedings of the European Design & Test Conference*, pages 328–333, Mar. 1996.

[9] T. R. Shiple. *Formal Analysis of Synchronous Circuits*. PhD thesis, UC Berkeley, Electronics Research Laboratory, College of Engineering, University of California, Berkeley, CA 94720, Oct. 1996. Memorandum No. UCB/ERL M96/76.

[10] G. Smith and R. Walford. The Identification of a Minimal Feedback Vertex Set of a Directed Graph. *IEEE Transactions on Circuits and Systems*, CAS-22(1):9–15, Jan. 1975.

[11] L. Stok. False Loops through Resource Sharing. In *Proceedings of the IEEE International Conference on Computer-Aided Design*, pages 345–348, Nov. 1992.

[12] R. v. Hanxleden, J. Bohne, L. Lavagno, and A. Sangiovanni-Vincentelli. Hardware/software co-design of a fault-tolerant communication protocol. In *Proceedings of the IEEE International Workshop on Embedded Fault-Tolerant Systems*, Dallas, Sept. 1996.

An Efficient Representation for Formal Synthesis *

Christian Blumenröhr, Dirk Eisenbiegler

Institute for Circuit Design and Fault Tolerance
(Prof. Dr.-Ing. D. Schmid)
University of Karlsruhe, Germany
blumen,eisen@ira.uka.de

Abstract

In the last years, performing synthesis by logical refinement has become an interesting alternative towards post-synthesis verification. This paper gives a case study about the complexity of formal synthesis programs within a given calculus. For a simple synthesis step, it is discussed, how one can efficiently implement circuit transformations with respect to the complexity of the logical transformations of the underlying calculus.

1. Introduction

Nowadays, more and more complex and sophisticated synthesis tools are involved during the design of digital circuits. However, even a fully automated synthesis process does not necessarily mean "correctness by construction". In general, bugs in synthesis tools lead to faulty implementations, and due to their complexity, it is almost impossible to formally verify synthesis tools. Therefore, formal methods have to be applied to ensure that the implementation satisfies the specification. A possible approach is post-synthesis verification [Gupt92]. However, fully automated post-synthesis verification is only achievable for comparatively small sized circuits at lower levels of abstraction, and theorem proving that requires interactive steps is not very much accepted by circuit designers.

Formal synthesis is a complementary approach to hardware verification, since formal averment is an integral part of the synthesis process. In [KBES96] a survey of and a classification scheme for formal synthesis approaches can be found. In the last years, formal synthesis has become a new research topic and several systems have been introduced such as T-Ruby [ShRa95], Lambda/Dialog [MaFo91], Veritas [HaLD89], DDD [JoBo91] or HASH [EiKB97]. [Busc92] and [BaFr96] present reimplementations of the Veritas formal synthesis approach for different theorem provers (LAMBDA and ISABELLE, respectively). Other research activities have been started by [GrMT94], [Wang92] or [Lars95]. They all have one thing in common: they are based on some calculus, i.e. some small core of basic logical transformations. The efficiency of formal synthesis approaches depends on how efficient hardware can be represented and on how fast circuit transformations can be realized based on the given set of logical transformations.

This paper investigates efficiency aspects of implementing formal synthesis transformations in the theorem proving environment HOL [GoMe93] illustrated by a basic circuit transformation step. In general, there are various ways to perform such transformations. However, the complexity very much depends on which basic logical transformations are used and on the order in which the logical transformations are applied. Therefore, when implementing such synthesis transformations, one has to consider the time complexity of the basic logical transformations involved. We will introduce a simple algorithm and discuss, how its performance can be improved.

In order to optimize the performance of formal synthesis, it is also possible to modify the core of the theorem prover such that the considered circuit transformations can be implemented more efficiently. However, one has to be careful, since modifying the core of a theorem prover may violate the consistency of the calculus. We will introduce a modification of the HOL theorem prover and discuss the impact towards safety and efficiency.

At the end of the paper, we will discuss the extra costs for formal synthesis as compared to conventional synthesis. In our approach HASH (Higher order logic Applied to Synthesis of Hardware), which exploits standard synthesis algorithms, circuit transformations are strictly divided from the design space exploration parts (calculating the scheduling table, determining the state encoding, etc.). The design

*This work has been partly financed by the Deutsche Forschungsgemeinschaft, Project SCHM 623/6-1.

1080-1820/97 $10.00 © 1997 IEEE

space exploration parts are the same as in conventional synthesis programs [TLWN90, GDWL94]. The extra cost for formal synthesis arise from the circuit transformations, that are to be performed by rule applications within a theorem prover.

2. Scheduling transformation

In this paper, we concentrate on the transformation in HASH for performing the scheduling task within high-level synthesis. High-level synthesis converts an algorithmic description of the circuit into a structure at the Register-Transfer (RT) level. The major steps in high-level synthesis are scheduling, allocation of storage, functional and interconnection units, binding the allocated hardware onto some library components and interface synthesis.

The scheduling task assigns a control step (c-step) to each operation in the algorithmic specification. There exist various heuristic scheduling algorithms which try to minimize the number of control steps or the hardware requirements [CaWo91, GDWL94]. A large number of them start from data flow graphs that correspond to the basic blocks in the algorithmic description. Although certain scheduling algorithms start from control/data flow graphs, we shall restrict ourselves to pure data flow graphs in this paper. The handling of control flow is the topic of our current research. In figure 1, the scheduling task is performed for a small data flow graph.

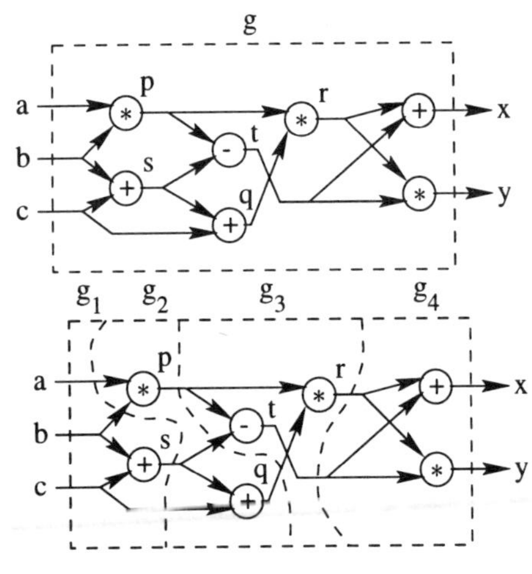

Figure 1. Example for a scheduling step

2.1. Formalizing data flow graphs

In order to transform the data flow graphs, they first have to be formalized suitably. In our approach, data flow graphs are represented by means of functions that are nothing but simple compositions of basic operations. They are formalized using λ-expressions [Davi89].

The following term shows, how the data flow graph in figure 1 is represented in HOL:

$$\begin{array}{l}\lambda(a,b,c).\\ \quad \mathsf{let}\ p = a*b\ \mathsf{in}\\ \quad \mathsf{let}\ s = b+c\ \mathsf{in}\\ \quad \mathsf{let}\ q = s+c\ \mathsf{in}\\ \quad \mathsf{let}\ r = p*q\ \mathsf{in}\\ \quad \mathsf{let}\ t = p-s\ \mathsf{in}\\ \quad \mathsf{let}\ x = r+t\ \mathsf{in}\\ \quad \mathsf{let}\ y = r*t\ \mathsf{in}\\ \quad\quad (x,y)\end{array}$$

The above expression describes an input/output function in terms of its basic operations. The function maps some input triple (a, b, c) to an output pair (x, y). Each let-term describes the connectivity of one operation. let-terms stand for β-redices, where $\mathsf{let}\, x = y\, \mathsf{in}\, z$ means $(\lambda x.z)\, y$. Using let-expressions improves readability.

2.2. Transforming the data flow graphs within HOL

During scheduling, the function g is split into a concatenation of functions $g_1, g_2, \ldots, g_k$ with $g = g_k \circ \ldots \circ g_2 \circ g_1$, and each function again represents a data flow graph (see figure 1).

This section describes, how the scheduling process described in figure 1 is implemented as a conversion in HOL. Our high level synthesis conversion is steered by external control information (the schedule table). The implementation of other high level synthesis conversions (allocation and binding of storage and (multi-purpose) functional units) is described in [EiBK96]. In this section we will only describe the logical aspects of formally deriving the synthesis result from the input data flow graph. The computation of the control information and invocation of the external heuristics will be discussed in section 4.

The approach is based on a conversion for normalizing functions. We will first describe this conversion and then describe, how scheduling can be realized based on this conversion.

2.3. Function normalization

The HOL representations corresponding to figure 1 are both nothing but simple compositions of the same basic functions. In principle, normalizing such representations is pretty simple. The general algorithm looks as follows:

1. the original term g is converted to $\lambda(x_1, x_2, \ldots x_m).g(x_1, x_2, \ldots x_m)$ by applying a paired η-reduction in the inverse direction

2. the $\circ$ operations are expanded by rewriting, provided there are any

3. β-reductions and paired β-reductions are performed wherever possible

Given some function g and the scheduled function $g' = g_k \circ \ldots \circ g_2 \circ g_1$, this algorithm leads to the same normalized representation, which looks like $\lambda(x_1, x_2, \ldots x_m).v[x_1, x_2, \ldots, x_m]$. In $v[x_1, x_2, \ldots x_m]$ there are no β-redices left and there is nothing but pure function applications.

The most significant step performed in this normalization, is (paired) β-reduction. β-reduction means turning some expression $(\lambda x.P[x])\, a$ to $P[a/x]$, where x is substituted by a in P. Paired β-reduction means turning some expression $(\lambda(x_1, ..., x_n).P[x_1, ..., x_n])\, (a_1, ..., a_n)$ to $P[a_1/x_1, , ..., a_n/x_n]$.

2.4. A universal conversion

We will now introduce a simple conversion which is based on this normalization scheme. Given some data flow graph representation g and some schedule table, the following steps have to be performed:

1. produce $g' = g_k \circ \ldots \circ g_2 \circ g_1$ according to the schedule table

2. derive $\vdash g = \hat{g}$ and $\vdash g' = \hat{g}$ via normalization

3. The equations $\vdash g = \hat{g}$ and $\vdash g' = \hat{g}$ are combined to $\vdash g = g'$ (symmetry and transitivity of equivalence).

The major drawback of this universal conversion is the complexity of step 2 when dealing with data flow graphs with a big depth, i.e. maximum number of operations on a path from some input to some output. Data flow graphs whose intermediate nodes have larger fanouts, i.e. the output of a node is used by many successor nodes as inputs, lead to a number of duplications during β-reduction. Due to the fact that during β-reduction, one has to traverse the entire term and since such β-redices can be nested, the term size and time consumption in step 2 may grow exponentially with the depth.

2.5. An advanced conversion

The universal conversion does not exploit any knowledge about how the synthesis step was performed. However, one can think of an advanced conversion, where synthesis is performed by a sequence of conversions which are optimized for the scheduling step.

In principle, this conversion is similar to the universal conversion except that step 2 is tuned towards scheduling. The idea of our scheduling conversion is to split the data flow graph step by step rather than doing it all at once, as in the universal synthesis conversion. β-reduction is only applied to those variables whose corresponding nodes have been assigned to the current control step. Although some β-redices will remain, the terms achieved after normalization will be equal.

Other than in the universal synthesis conversion, $k-1$ conversions (k – number of control steps) have to be applied successively rather than applying one single conversion. Hence, the exponential complexity associated with step 2 is avoided and the overall cost is reduced.

To demonstrate the differences between the simple and the advanced conversion, we will show some experimental results in the following section.

2.6. Experimental results

We consider two scalable data flow graphs. As a first example, we use a data flow graph, which realizes the division of two polynomials (PD) with the given coefficients α_i and β_i:

$$\frac{\sum_{i=0}^{p+q} \alpha_i\, x^i}{\sum_{i=0}^{p} \beta_i\, x^i} = \sum_{i=0}^{q} \gamma_i\, x^i + \frac{\sum_{i=0}^{p-1} \delta_i\, x^i}{\sum_{i=0}^{p} \beta_i\, x^i}$$

The coefficients γ_i and δ_i should be computed. To facilitate the calculation, we assume that the divisor is normalized with respect to β_p. After a few algebraic transformations we get the following two formulas for computing the coefficients:

$$\begin{aligned} \gamma_i &= \alpha_{i+p} - \sum_{k=i+1}^{min\{i+p,q\}} \beta_{i+p-k} \cdot \gamma_k & i = 0 \ldots q \\ \delta_j &= \alpha_j - \sum_{k=0}^{min\{j,q\}} \beta_{j-k} \cdot \gamma_k & j = 0 \ldots p-1 \end{aligned}$$

The data flow graph consists of $p+q$ subtractors, $p(q+1)$ multipliers and $q(p-1)$ adders, so there is a total of $2pq+2p$ nodes. The critical path has a length of $3q+2$ nodes.

Another scalable data flow graph is realized in our second example. It calculates the discrete cosine transform (DCT), which is popularly used for image compression. The DCT of an image with pixels $x(n,m)$ is defined by:

$$X(u,v) = \frac{2}{\sqrt{N \cdot M}} \cdot c(u) \cdot c(v) \cdot \sum_{n=0}^{N-1} \sum_{m=0}^{M-1} x(n,m) \cdot cos[\frac{\pi \cdot u}{2N} \cdot (2n+1)] \cdot cos[\frac{\pi \cdot v}{2M} \cdot (2m+1)]$$

with

$$c(u), c(v) = \begin{cases} \frac{1}{\sqrt{2}} & : \quad u, v = 0 \\ 1 & : \quad \text{otherwise} \end{cases}$$

In the following, we assume that $N = M$. In most applications, the parameters are set to $N = M = 8$. The number of additions is $2N^3 - N^2 - N$ and there are $2N^3 - N + 2$ multiplications. So there is a total of $4N^3 - N^2 - 2N + 2$ nodes. The length of the critical path is $2N + 1$. A more detailed description of this data flow graph can be found in [BlEK96].

In figure 2, the runtimes for the simple and advanced conversion are shown for the polynomial division. We set p to 25 and increased q. Due to the exponential memory consumption of the simple conversion, the computer's capacity of 1.2 GB was exceeded at about 600 nodes.

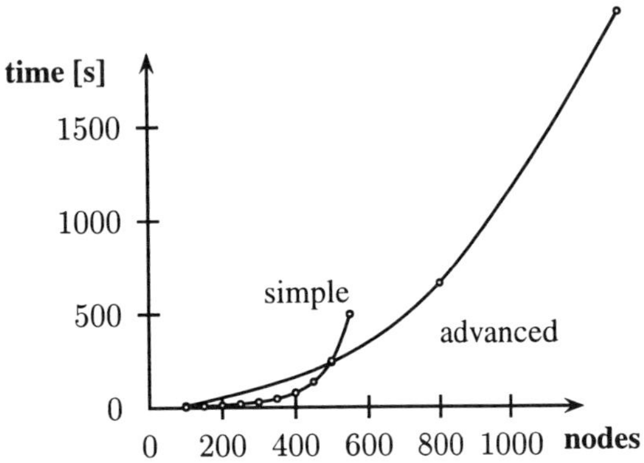

Figure 2. Simple and advanced conversion applied to PD

Figure 3 shows the runtimes for the two above mentioned data flow graphs. We applied different schedules that were derived by different scheduling algorithms namely ALAP and FD (force-directed [PaKn89]). For PD, p was again set to 25, and for the DCT, we increased N from 2 to 8. The structures of the data flow graphs PD and DCT differ in the depth and the number of reused intermediate results. As can be seen, the runtime depends on the structure of the DFG but is almost independent of the schedule table.

3. Changing the core of the theorem prover

In the following, we will present a modified HOL theorem proving system, named HOL', where we changed the core in order to increase efficiency. There are two modifications: the term representation is changed and also the core of basic logical transformations is enlarged.

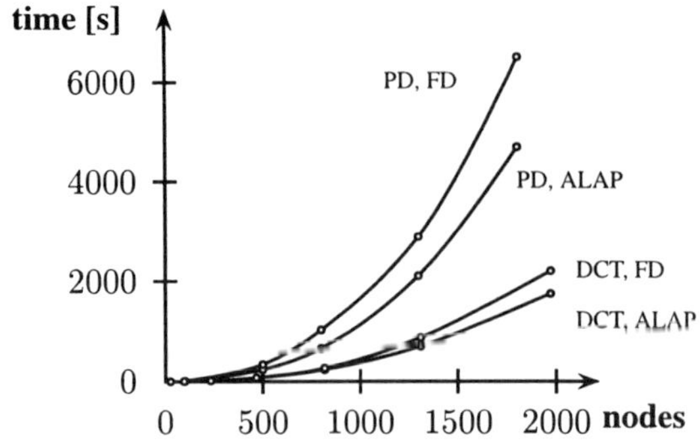

Figure 3. Advanced conversion for different data flow graphs and schedules

3.1. Changing the term representation

In the HOL theorem prover, terms are implemented in a so called *deBrujin*-style, which means that free and bound variables are represented differently. Free variables are stored with their name and type, whereas bound variables are only represented by a number, linking to the λ-abstraction it refers to. The number tells, how many λ-abstractions apart the referred λ-abstraction stands. Example: The variables x and y in the body of the term $\lambda x.(\lambda y.x\,y)$ are internally represented by 1 and 0, respectively. One advantage of the deBrujin representation is, that checking the α-equivalence of two terms can be performed in constant time.

However, the deBrujin term representation has a decisive drawback. Whenever one traverses a term while constructing or destructing it and reaches a bound variable, one has to go back to find the corresponding λ-abstraction for identifying the variable. Especially, if, as in our approach, variables are bound over a large range, this leads to a quadratic complexity. Therefore deBrujin term representation is not well suited for our application. However, this problem does not only arise in our formal synthesis approach. Whenever large circuits have to be represented, variables, which correspond to signals, are bound over a large range, and this means that deBrujin term representation becomes very inefficient.

Therefore we implemented HOL with a so called *name-carrying* term representation. Here, also the bound variables are stored with their name and type. It has the advantage that terms, which correspond to circuit descriptions, can be handled in a by far more efficient manner. On the other hand, some basic operations (e.g. α-equivalence check) become a bit more inefficient when using this representation style. Furthermore, it also increases memory consumption for representing bound variables, since instead of just a number, a string (for the name) and a type expression have to be stored.

3.2. Introduction of more efficient functions

As mentioned in section 2.3, β-reduction is applied very often during the normalization of the terms. HOL only allows one single β-reduction at a time. However one could increase the efficiency of the scheduling transformation by performing several β-reductions in a single term traversal step. By adding this conversion to the core, we improve the efficiency of the theorem prover. This advanced β-conversion is equipped with a filter for selecting the bound variables that are to be expanded. This is important for our advanced conversion, where we only expand variables, that occur within the considered control step (see section 2.5). Based on the advanced β-conversion, paired β-reduction can be performed within one traversal. For the implementation of the advanced paired β-conversion we added a conversion, which changes paired β-redices into an unpaired representation. Example: $(\lambda(x, y).t)\ (a, b)$ is turned to $(\lambda x.(\lambda y.t))\, a\, b$. Afterwards, all the β-redices can be expanded in one traversal using the already mentioned advanced β-conversion.

The additional conversions are not only tailored to our application, but is of general interest for other users of the HOL system. However, it is to be noted, that enlarging the core is safety critical. Therefore, one has to consider carefully, which functions should be added to the core.

Figure 4 shows runtimes for a scheduling of the PD data flow graph scheduled with an ALAP technique. It shows, that both the simple and the advanced conversion runs faster under the modified HOL system (indicated by HOL'). Again, due to nested β-redices, the simple conversion ran into memory problems for larger data flow graphs.

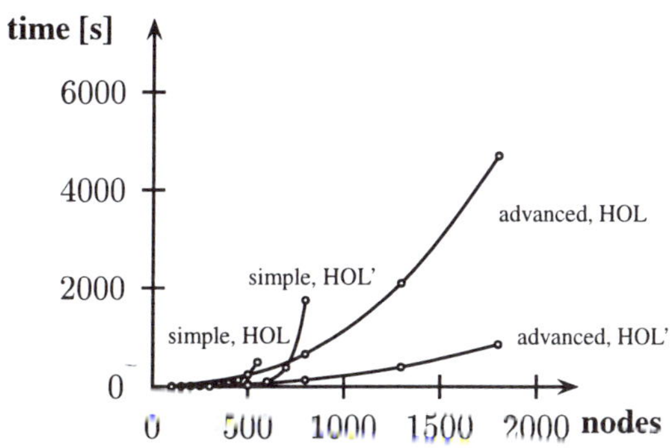

Figure 4. Simple and advanced conversion in different theorem provers

4. The formal synthesis scenario

Figure 5 demonstrates the underlying idea of our formal synthesis approach HASH illustrated by the scheduling step. Given a data flow graph, some scheduling heuristic is started. This heuristic step has nothing to do with logic. The heuristic returns a scheduling table which maps each operation in the data flow graph onto a control step. This scheduling table is now used by the formal logical transformation in HASH to produce a scheduled data flow graph.

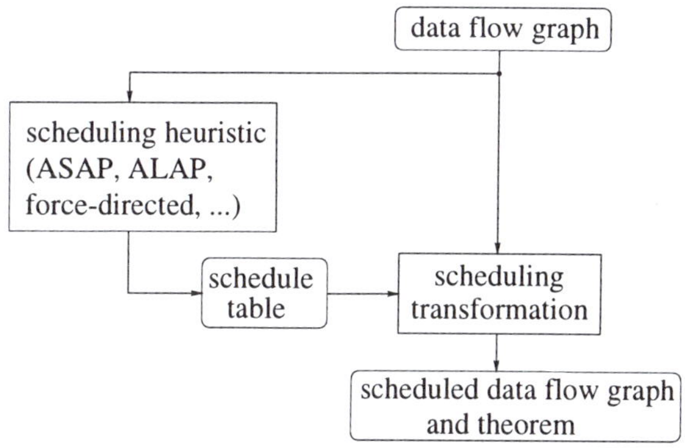

Figure 5. Invoking synthesis heuristics within formal synthesis

The split between design space exploration (i.e. different schedule tables for different heuristics) and the logical transformation is the core idea in HASH. This core idea is applicable to most of the synthesis steps, e.g. allocation/no. of resources available, retiming/split in the combinational logic, etc.

Two important points are met independently with this strategy: quality and correctness of the implementation. The quality only depends on the algorithm that calculates the control information, whereas the correctness aspect is guaranteed due to the transformation being based on the HOL system.

Since the entire synthesis process is nothing but a HOL conversion, correctness is guaranteed implicitly. Faulty implementations *cannot* be achieved even if the control information produced by the external program is flawed, such as a schedule where the data dependencies are disregarded. In such cases, the transformation cannot be performed within the logic and an exception will be raised. In conventional synthesis programs, such bugs could lead to faulty implementations. Our formal synthesis program either leads to correct implementations or to no implementation but an exception. In case of an exception, an information is produced telling the user in which synthesis step the error occurred.

In our approach, circuit transformations guarantee that the functional behavior is preserved. Besides functional correctness, there may be further requirements for the implementation such as timing and area constraints. However, checking, whether such constraints are fulfilled, can easily be done by non-formal methods. Therefore, it is not necessary to formally represent and verify such constraints in logic. In our scheduling step, for example, it is pretty easy to count the number of control steps and to check, whether it exceeds some given limit.

Figures 6 and 7 show the runtimes for both heuristics and transformational part of the formal synthesis step for the PD data flow graph and the DCT data flow graph, respectively (notice the different scales!). For determining the schedule table, we applied both the force-directed and the ALAP program. The circuit transformation was performed in HOL' (the modified HOL system) by applying the advanced conversion. As can be seen, the runtime for sophisticated design space exploration techniques such as force-directed scheduling can exceed the runtime for the transformational part by far. Even for simple design space exploration techniques, the ratio between design space exploration and circuit transformation seems reasonable.

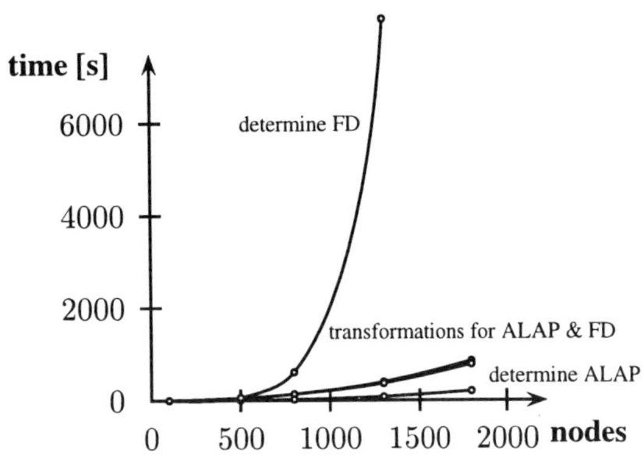

Figure 6. Time consumption for heuristic and transformation for PD

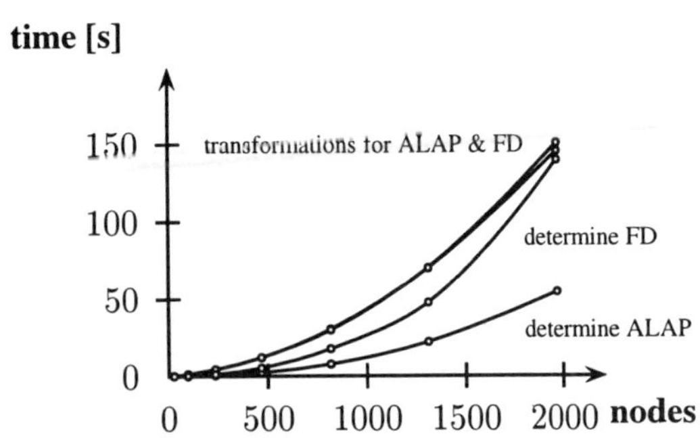

Figure 7. Time consumption for heuristic and transformation for DCT

5. Conclusions

For a given simple synthesis step, we have illustrated the efficiency problem when performing the step by means of a logical transformation. It shows, that it is very important to be aware of the complexity of the basic logical transformation when constructing formal synthesis programs. On the other hand it can be worthwhile to modify the basic logical transformations themselves. We have demonstrated, that formal refinement techniques for hardware synthesis can only be applicable in practice when considering these efficiency aspects. Efficient formal synthesis implementations, however, are applicable even for large sized circuits and the extra-costs for it, which are independent of the design space exploration part, are reasonable.

References

[BaFr96] D. Basin and S. Friedrich. Modeling a hardware synthesis methodology in isabelle. In [HOL96], pages 33–50.

[BlEK96] C.Blumenröhr, D. Eisenbiegler, and R.Kumar. Applicability of formal synthesis illustrated via scheduling. In *Workshop on Logic and Architecture Synthesis*, Grenoble, France, December 1996. Institut National Polytechnique de Grenoble.

[Busc92] H. Busch. Transformational design in a theorem prover. In V. Stavridou, T. F. Melham, and R. T. Boute, editors, *Theorem Provers in Circuit Design*, volume A-10, pages 175–196, Nijmegen, The Netherlands, June 1992. IFIP TC10/WG10.2 International Conference, North-Holland.

[CaWo91] R. Camposano and W. Wolf. *High-Level VLSI Synthesis*. Kluwer, Boston, 1991.

[Davi89] R. E. Davis. *Truth, Deduction and Computation: Logic and Semantics for Computer Science*. Computer Science Press, New York, 1 edition, 1989.

[EiBK96] D. Eisenbiegler, C. Blumenröhr, and R. Kumar. Implementation issues about the embedding of existing high level synthesis algorithms in HOL. In [HOL96], pages 157–172.

[EiKB97] D. Eisenbiegler and R. Kumar and C. Blumenröhr . A constructive approach towards correctness of synthesis-application within retiming. In *The European Design & Test Conference*, pages 427–432, Paris, France,

March 1997. IEEE Computer Society and ACM/SIGDA, IEEE Computer Society Press.

[GDWL94] D. Gajski, N. Dutt, A. Wu, and S. Lin. *High-Level Synthesis, Introduction to Chip and System Design*. Kluwer Academic Publishers, 1994.

[GoMe93] M.J.C. Gordon and T.F. Melham. *Introduction to HOL: A Theorem Proving Environment for Higher Order Logic*. Cambridge University Press, 1993.

[GrMT94] W. Grass, M. Mutz, and W. Tiedemann. High level synthesis based on formal methods. In *Proc. EUROMICRO*, pages 83–91, Liverpool, 1994.

[Gupt92] A. Gupta. Formal hardware verification methods: A survey. *Formal Methods in System Design*, 1(2/3):151–238, 1992.

[HaLD89] F.K. Hanna, M. Longley, and N. Daeche. Formal synthesis of digital systems. In Luc J. M. Claesen, editor, *Applied Formal Methods For Correct VLSI Design*, volume 2, pages 532–548. IMEC-IFIP, Elsevier Science Publishers, 1989.

[HOL96] Joakim von Wright, Jim Grundy, and John Harrison, editors. *Theorem Proving in Higher Order Logics:9th International Conference, TPHOLs'96*, number 1125 in Lecture Notes in Computer Science, Turku,Finland, August 1996. Springer-Verlag.

[JoBo91] S.D. Johnson and B. Bose. DDD - A system for mechanized digital design derivation. In *Workshop on Formal Methods in VLSI Design*, Miami, Florida, January 1991. ACM/SIGDA.

[KBES96] R. Kumar , C. Blumenröhr, D. Eisenbiegler, and D. Schmid . Formal synthesis in circuit design-A classification and survey. In M. Srivas and A. Camilleri, editors, *Formal Methods in Computer-Aided Design. First International Conference,FMCAD'96*, number 1166 in Lecture Notes in Computer Science, pages 294–309, Palo Alto, CA, USA, November 1996. Springer-Verlag.

[Lars95] M. Larsson. An engineering approach to formal digital system design. *The Computer Journal*, 38(2):101–110, 1995.

[MaFo91] E.M. Mayger and M.P. Fourman. Integration of formal methods with system design. In A. Halaas and P.B. Denyer, editors, *International Conference on Very Large Scale Integration*, pages 59–70, Edinburgh, Scotland, August 1991. IFIP Transactions, North-Holland.

[PaKn89] P. G. Paulin and J. P. Knight. Force-directed scheduling for the behavioral synthesis of ASIC's. *IEEE Transactions on Computer Aided Design*, 8(6):661–679, June 1989.

[ShRa95] R. Sharp and O. Rasmussen. The T-Ruby design system. In *CHDL '95*, pages 587–596, 1995.

[TLWN90] D.E. Thomas, E.D. Langnese, R.A. Walker, J.A. Nestor, J.V. Rajan, and R.L. Blackburn. *Algorithmic and Register-Transfer Level Synthesis: The System Architect's Workbench*. Kluwer Academic Publishers, 1990.

[Wang92] L. Wang. Deriving a correct computer. In L. J. M. Claesen and M. J. C. Gordon, editors, *Higher Order Logic Theorem Proving and its Applications*, volume A-20, pages 449–458, Leuven, Belgium, September 1992. IFIP TC10/WG10.2, North-Holland.

Derivation of Formal Representations from Process-based Specification and Implementation Models

Steven Vercauteren*, Diederik Verkest*, Gjalt de Jong† and Bill Lin‡
* IMEC Laboratory, Kapeldreef 75, B-3001 Leuven, Belgium
† Alcatel Telecom, F.Wellesplein 1, B-2018 Antwerpen, Belgium
‡ ECE Department, University of California, San Diego, La Jolla, CA

Abstract

In this paper we present a formal framework to verify timing properties of embedded systems. We propose a process calculus as an intermediate model to map between language-level constructs of process-based specification and implementation models, and Petri net operations. We present an elegant translation scheme to generate Petri nets starting from the intermediate process expressions. The approach has been applied to verify the freedom of deadlock in a QAM modem design, with promising results.

1. Introduction

As embedded systems become increasingly complex, they become increasingly prone to errors and difficult to design. From experience gained with several large scale design projects in digital communications [4, 8], we have encountered a number of subtle errors during both the specification and implementation phases, that were difficult to trace using traditional analysis methods such as simulation. Since most modern embedded systems are concurrent in nature, usually implemented using a heterogeneous architecture containing multiple hardware/software components, hard to detect errors can result from unanticipated interactions between the concurrent parts. Over the past years, a number of automated formal verification methods [5, 6, 12, 15] have emerged, that can address some of these verification problems. However, to leverage these methods, the specification model used to represent the system must be formal in the first place as ambiguity in specification semantics is in itself a major source of errors.

In this paper, we propose a formal model based on Petri nets [13] for reasoning about the behavior of a concurrent system. We have chosen the Petri net formalism, because it is well suited to model concurrency, choice, and causality, and because there is a wealth of formal verification techniques [15, 18] that can exploit the inherent partial order properties explicitly captured in the model. However, a major problem when using Petri nets as a model for verification, is the lack of a formal mapping between constructs in the (system) specification language itself and Petri net operations. We address this problem by providing such a formal mapping using a CSP [9]/CCS [10]-like process calculus as an intermediate model. The underlying idea is that many of the existing process-based specification languages (e.g. [2]) can be translated to what we call "intermediate code" of this process calculus, which can subsequently be translated into a Petri net representation by applying a syntax-directed mapping scheme. Once arrived at the Petri net level, existing formal verification methods can then be leveraged. Because the overall mapping is semantics preserving, these methods can be used to validate the various intermediate specifications of the design process.

The impact of this work on the design of embedded systems is twofold. At the specification phase, errors can be detected quickly without requiring extensive simulations. Secondly, when designing an embedded system, many problems occur at the hardware/software interface. Therefore, a design support tool has been developed [19] to assist the designer in the mapping of software components on programmable processors. Such a tool assembles the hardware/software interface by selecting and combining I/O scenarios from a library. Individual scenarios are correct by construction. However, when assembling these scenarios to construct the complete interface, subtle (timing) problems may arise, as will be shown in section 5. The formal framework presented in this paper allows to diagnose these problems, such as possible deadlocks. This necessitates the modeling of the complete hardware/software interface, including the processor core, run-time operating system, hardware and software device drivers, etc.

The remainder of this paper is organized as follows. Section 2 presents the basic concepts of the used process calculus. Section 3 reviews the basic definitions of Petri nets, and details the *syntax-directed* translation procedure of process expressions to Petri nets. Section 4 highlights the verification framework. Section 5 extensively describes a case study. Finally, conclusions are drawn in section 6.

2 Process Calculus

In our model we start from a set of given **atomic rendez-vous actions**. These actions are taken to be indivisible, and form the "leaf" processes of the model. The occurrence of an action is called an event [1], and it is the result of two concurrent processes - in some cases the environment is implicitly assumed to be the counterpart process - both engaging in the execution of the event under consideration. Every event can thus be seen as a **binary** synchronization. Two types of events exist: *visible* events and *hidden* events. Visible events represent *external* synchronizations (i.e. which are externally observable); the set of all visible events is denoted as $\boldsymbol{Vis}$. Hidden events represent *internal synchronizations* and are denoted by the special symbol τ. The environment cannot prevent a process from engaging in a *hidden* event. Indeed, as τ denotes an internal synchronization and all synchronizations are binary, it cannot be synchronized upon by any other process.

Processes are built from other (leaf) processes by means of the following operators: *sequential composition*, *deterministic* and *non-deterministic choice composition*, *parallel composition*, *recursion* and *interrupt composition*.

Processes are denoted by *process expressions*. We also assume a set $\boldsymbol{Var}$ of *process variables* denoted by x, y, etc. The set $\boldsymbol{PE}$ of all possible process expressions is defined by the following production system, where R is used as start symbol:

$$R ::= S \mid T$$
$$S ::= x \mid S.R$$
$$T ::= a \mid T.R \mid T + T \mid T^\circ \| T^\circ \mid T^\circ \rhd T^\circ \mid \mu x.T$$
where $a \in (\boldsymbol{Vis} \cup \{\tau\})$
T° is T without free occurrences of variables

In the following, the intuitive meaning of the different operators is described.

Sequential Composition of two processes P and Q, denoted $P.Q$, is a process that starts the execution of Q after the *successful completion* of P. Successful completion of a process P is defined as process P reaching a wanted [2] state from which no actions can be performed. This operator constitutes a mixture of the *prefix* and *sequential composition* operators, as they are used in CSP [9].

Choice. The deterministic choice of two processes P and Q, denoted $P + Q$, is the process that can do either P or Q, where the choice is made by the environment. The non-deterministic choice of two processes P and Q, on the other hand, denoted $P \sqcap Q$, is the process that can either do P or Q, where the choice is made by the process itself. In both cases, the decision of which alternative to take, should be made at the beginning of the choice. Notice that the notation $P \sqcap Q$ is a shorthand for the process $(\tau.P) + (\tau.Q)$. Indeed, as τ denotes a hidden or internal synchronization event, the outside world cannot interfere at the beginning of the choice; $(\tau.P) + (\tau.Q)$ then degenerates to a non-deterministic choice. This construction is borrowed from CCS [10].

Parallel Composition of two processes P and Q, denoted $P\|Q$, executes the processes P and Q simultaneously and independently, except for events that are common to both processes; if both processes contain an event a, this event can only occur if both parties are ready to engage in a. With the exception that we only allow binary synchronization, this operator is also used in CSP.

Interrupt Composition of process P by a process Q, denoted as $P \rhd Q$, is the process that behaves like P but which is *interrupted* on the occurrence of the first event of Q. When interrupted, process P arrives in a *halt* state, and $P \rhd Q$ behaves like Q. After Q has terminated successfully, P leaves its halt state and $P \rhd Q$ resumes P, until possibly being interrupted again. If P terminates unsuccessfully, it is still possible for $P \rhd Q$ to behave like Q. If P terminates successfully, on the other hand, Q cannot start execution any more. The semantics of the interrupt composition, as they are presented here, differ from the ones defined by CSP and LOTOS [11]. In the latter models, an interrupt is interpreted as an *abort*; when a process Q "interrupts" a process P, process Q is executed, but P is never resumed. Clearly, this kind of preemption cannot be reduced to the interrupt behavior of today's processors.

Recursion is introduced by considering solutions of equations of the form $x = F(x)$, denoted as $\mu x.F(x)$, with $F(x)$ an expression in (process) variable x, constructed solely in terms of the sequential composition, choice composition, parallel composition and interrupt composition operators. Notice that an expression beginning with a variable, can only be used in a composition if it is preceded by an event $a \in \boldsymbol{Vis} \cup \{\tau\}$.

Besides the interrupt operator, our process calculus has no larger expressiveness than CSP and CCS. The key idea of our approach is that we believe many of the existing process-based languages (e.g. [2]) used to specify heterogeneous systems, can be translated to "intermediate code" of this calculus which then serves as a starting point for performing a syntax-directed Petri net translation, as described in the next section.

Example. Consider the following program fragment, written in C, as it can be found in a process description within the CoWare™data model [2]:

```
while(!(x%y)) {
 send(a,x);
 x++;
 receive(b,y);
}
 send(c,(x+y));
```

The send(a,x) statement initiates a communication to send data along channel a. If the environment, at the other

[1] In the sequel we will indistinctively use the notations "event" and "action"

[2] Because we have to exclude the deadlock state.

side of the channel, is not ready to receive data along this channel, the program is halted at this point. As soon as the environment becomes ready, data is transfered, and both parties proceed independently. The occurrence of this communication is modeled by the atomic event a. For the `receive(b,y)` and `send(c,(x+y))` statements the same reasoning applies. This program fragment is then modeled as $X = (a.b.X) \sqcap c$. Every time we encounter the `while` statement, the decision of (re-)entering the loop is made internally `(!(x%y))` - we abstract from all internal data -and cannot be influenced by the environment. As a result, it is modeled as a non-deterministic choice.

3. Syntax-directed Translation to Petri Nets

In this section the translation of process expressions into Petri nets [13] is described in more detail. In section 3.1 we review some of the basic definitions and properties of Petri nets. In section 3.2 the composition operators are defined at the Petri net level. In section 3.3 we discuss the translation process itself.

3.1. Basic Definitions and Properties

Definition 3.1 (Labeled Petri Net) *A labeled Petri net (LPN) is a tuple* $\Sigma = \langle P, A, F, m_0 \rangle$ *with* $P \cap A = \emptyset$, $F \subseteq 2^P \times A \times 2^P$ *and* $m_0 : P \to \mathbb{N}$.

In the above definition P denotes a set of *places*, A a set of *actions*, F the set of *transitions* and m_0 an initial marking. For a transition $t = (p, a, q)$, a denotes the *label* or the *action* of t, where as p and q are often referred to as the set of *input* and *output* places of t, respectively.

Besides the structure of a Petri nets, there is also an associated dynamics. A *state* or marking, is the mapping of the places to the natural numbers, indicating the number of tokens in the places. Transitions between states are dictated by the following firing rule. In the sequel M_P denotes the set of all states (markings) of a Petri net with $|P|$ places.

Definition 3.2 (Enabling Rule) *Let* (p, a, q) *be a transition,* $m \in M_P$. *Enabled*$((p, a, q), m) \equiv \forall p : m(p) \geq 1$

Definition 3.3 (Firing Rule) *Let* $t = (p, a, q)$ *be a transition,* $m \in M_P$, *Enabled*(t, m) = *true.*

$$NextState(m(p'), t) = \begin{cases} m(p') - 1 & \text{if } p' \in p \setminus q \\ m(p') + 1 & \text{if } p' \in q \setminus p \\ m(p') & \text{otherwise} \end{cases}$$

Definition 3.2 states that a transition t can fire if all its input places contain at least one token. Definition 3.3 states that firing of t removes one token in all its input places and adds a new token in all its output places.

The set of all reachable states is represented in a *reachability graph*. In such a reachability graph all vertices correspond to a valid marking of the Petri net and all arcs correspond to a transition from one marking to another due to firing of some transition in the net. The reachability graph of a Petri net N, denoted as $RG(N)$, can then be interpreted as the reflexive transitive closure of the next-state relation defined in definition 3.3.

Two important properties of Petri nets are *liveness* and *safeness*. Liveness concerns the question whether a transition can ever be fired, and is clearly opposed to *deadlock*. Safeness means that a place *does not* contain more than one token at any time.

3.2. Process Operators on Petri Nets

In this section the process of translating process expressions into Petri nets is highlighted. The approach is similar to the macro-module mapping approach [16] for translating a concurrent program into an asynchronous circuit. For each syntactical construct a Petri net element or operator is defined. In this way a process expression can be translated into a Petri net using a *syntax-directed mapping* scheme, detailed below.

In literature a lot of research has been devoted to the development of operators for the composition of Petri nets. In [17] finite and safe nets are constructed for the *Algebra of Communicating Processes*(ACP) of Bergstra and Klop [1] without recursion. The notion of a non-deterministic choice is absent either. In [14] safe and finite Petri nets are generated from a so-called *anonymous* language that contains CCS and almost the whole CSP as special "cases". The alternatives of a choice may however not contain parallelism, and every body of a recursion starts with an invisible action τ. In [7] a *Communication Petri Net Model* is proposed. The focus is a Petri net algebra; the concepts of successful completion (cf. sequential composition) and recursive processes are therefore missing.

In this work, the translation of a process expression into Petri nets is defined by means of translation function $\mathcal{PT}$. The syntax-directed organization of $\mathcal{PT}$ makes it necessary to include all possible syntactic constructs of a process expression in the domain of $\mathcal{PT}$, i.e. the set of variables $\boldsymbol{Var}$, the atomic actions, as well as the process operators.

As discussed in section 2, the sequential composition operator requires the notion of successful completion. As a result, the definition of a Petri net has to be extended to represent successful termination. Therefore we have chosen to classify certain places of the Petri net as *end places*. When all end places contain a token, the involved Petri net has terminated successfully. This new Petri net model is denoted as a *PE-net* and is defined below.

Definition 3.4 *A PE-net is a tuple* $\Sigma = \langle P, T, A, m_0, E \rangle$, $P \cap T = \emptyset$, $F \subseteq 2^P \times A \times 2^P$, $m_0 \subseteq P$ *and* $E \subseteq P$.

In the above definition P, T, F and m_0 have the same meaning as in (classical) labeled Petri nets (see definition 3.1) and E denotes the end places. The definitions and properties of (classical) labeled Petri nets can be lifted to

PE-nets in a straightforward way. The initial marking of a PE-net, however, is represented by the set of *initial places*, i.e. the set of places that contain a token in the initial marking. This is possible due to the safe Petri net representations of the allowed process expressions. A safe Petri has in each place at most one token and its marking can therefore be represented by a set of places m, where $p_i \in m$ indicates that there is a token in p_i. A non-safe representation would imply that the corresponding process expression exhibits *auto-concurrency*, i.e. is of the form $x = P\|x; \ldots$, which is not allowed in our syntax.

Leaf Processes. The PE-net representations of an atomic action $a \in \boldsymbol{Vis} \cup \{\tau\}$ and a variable $x \in \boldsymbol{Var}$ are illustrated in Figure 1. The end places are depicted as circles with double perimeter.

Figure 1: PE-net representations of (a) $a \in \boldsymbol{Vis} \cup \{\tau\}$ (b) $x \in \boldsymbol{Var}$

Definition 3.5 *Given an atomic action $a \in \boldsymbol{Vis} \cup \{\tau\}$, the corresponding PE-net, denoted as $\mathcal{PT}(a)$, is defined as* $\langle\{p_1, p_2\}, \{a\}, \{(\{p_1\}, a, \{p_2\})\}, \{p_1\}, \{p_2\}\rangle$

Definition 3.6 *Given a variable $x \in \boldsymbol{Var}$, the corresponding PE-net, denoted as $\mathcal{PT}(x)$, is defined as* $\langle\{p_1, p_2\}, \{x\}, \{(\{p_1\}, x, \emptyset)\}, \{p_1\}, \{p_2\}\rangle$

Sequential Composition. For the PE-net representation of a process expression $Q_1.Q_2$, the end places of $\mathcal{PT}(Q_1)$ are combined with the initial places of $\mathcal{PT}(Q_2)$, by means of a Cartesian product, effectively "abutting" the two PE-net representations together. This is shown in Figure 2.

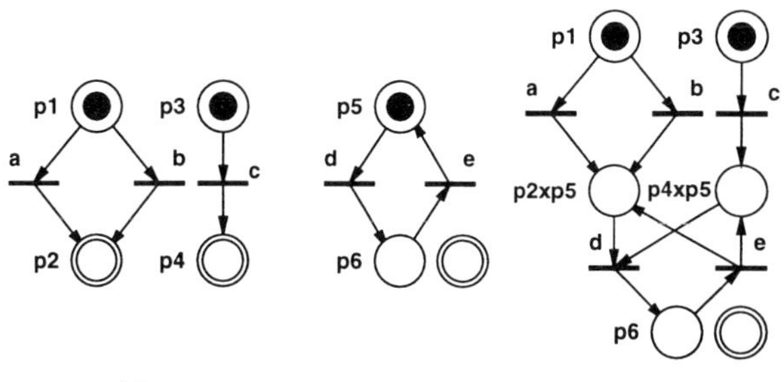

Figure 2: Petri net representations (a) $\mathcal{PT}(Q_1)$ (b) $\mathcal{PT}(Q_2)$ (c) $\mathcal{PT}(Q_1.Q_2)$

Definition 3.7 (Sequential Composition) *Let Q_1 and Q_2 be two process expressions, with $\mathcal{PT}(Q_1) = \langle P_1, A_1, F_1, M_{0_1}, E_1 \rangle$ and $\mathcal{PT}(Q_2) = \langle P_2, A_2, F_2, M_{0_2}, E_2 \rangle$. $\mathcal{PT}(Q_1.Q_2)$ is defined as:*
$\langle (P_1 \setminus E_1) \cup (P_2 \setminus M_{0_2}) \cup (E_1 \times M_{0_2}), A_1 \cup A_2, F', M_{0_1}, E_2 \rangle$
where

$$\begin{aligned} F' = \; & \{(p, a, q \setminus E_1 \cup (q \cap E_1) \times M_{0_2}) \mid (p,a,q) \in F_1\} \\ & \cup \{(p \setminus M_{0_2} \cup E_1 \times (p \cap M_{0_2}), a, \\ & \quad q \setminus M_{0_2} \cup E_1 \times (q \cap M_{0_2})) \mid (p,a,q) \in F_2\} \end{aligned}$$

In contrast to [17] there is no need for one-step unfolding of $\mathcal{PT}(Q_2)$, as well as an expensive *complement* step. In [14] sequential composition is defined as a special case of parallel composition; by doing so, however, there is a need for an extra *concealment* or *hide* operator.

Choice Composition. For the PE-net representation of a process $Q_1 + Q_2$, conflicts are introduced between all pairs of initial transitions of $\mathcal{PT}(Q_1)$ and $\mathcal{PT}(Q_2)$, by means of a Cartesian product construction. The end places are combined similarly. For this to work properly, the initial places that are in cycles have to extracted by a one-step unfolding; in a choice, once the decision of what branch to take is made by the first execution of a transition, a loop iteration may then not cause the other branch to be taken. In our translation scheme, the PE-net representation of a recursive process definition already implements the desired unfolding (see Definition 3.9). Then, the above situation can only result from process expressions of the form $P + (R \rhd Q) \ldots$. In these situations the one-step unfolding is accomplished through a separate *root-unwinding* step. This preprocessing step is only effective for those initial transitions that are in cycles. The PE-net equivalent of the choice operator is shown in Figure 3.

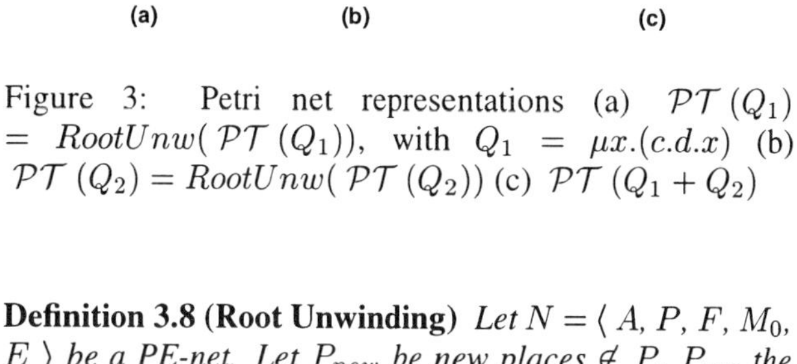

Figure 3: Petri net representations (a) $\mathcal{PT}(Q_1) = RootUnw(\mathcal{PT}(Q_1))$, with $Q_1 = \mu x.(c.d.x)$ (b) $\mathcal{PT}(Q_2) = RootUnw(\mathcal{PT}(Q_2))$ (c) $\mathcal{PT}(Q_1 + Q_2)$

Definition 3.8 (Root Unwinding) *Let $N = \langle A, P, F, M_0, E \rangle$ be a PE-net. Let P_{new} be new places $\notin P$, P_{cyc} the initial places of N, that are not in cycles, defined by $\{s \in P \mid M_0(s) \neq 0 \wedge \neg(\exists (p,a,q) \in F : s \in q)\}$, and η a bijection between P_{new} and P_{cyc}. The* root-unwinding *of N, denoted as $RootUnw(N)$ is defined as:*
$\langle A, P \cup P_{new}, F', (P \setminus P_{cyc}) \cup P_{new}, E \rangle$
where

$$F' = F \cup \{(p \cup p_n, a, q) \mid \ p_n \subseteq P_{new} \wedge (p \cup \mathrm{H}(p_n), a, q) \in F\}$$
$$\mathrm{H}(\{p_1, ..., p_n\}) = \{\eta(p_1), ..., \eta(p_n)\}$$

Definition 3.9 (Choice Composition) *Let Q_1 and Q_2 be two process expressions, with $RootUnw(\mathcal{PT}(Q_1)) = \langle P_1, A_1, F_1, M_{0_1}, E_1\rangle$ and $RootUnw(\mathcal{PT}(Q_2)) = \langle P_2, A_2, F_2, M_{0_2}, E_2\rangle$. $\mathcal{PT}(Q_1 + Q_2)$ is defined as:*

$$\langle((P_1 \setminus M_{01}) \setminus E_1) \cup ((P_2 \setminus M_{02}) \setminus E_2) \cup (M_{01} \times M_{02}) \cup E_1 \times E_2, A_1 \cup A_2, F', M_{01} \times M_{02}, E_1 \times E_2\rangle$$

where

$$\begin{aligned} F' = \ & \{(p_1 \setminus M_{01} \cup (p_1 \cap M_{01}) \times M_{02}, a_1, \\ & \quad q_1 \setminus E_1 \cup (q_1 \cap E_1) \times E_2) \mid (p_1, a_1, q_1) \in F_1\} \\ & \cup \{(p_2 \setminus M_{02} \cup M_{01} \times (p_2 \cap M_{02}), a_2, \\ & \quad q_2 \setminus E_2 \cup E_1 \times (q_2 \cap E_2)) \mid (p_2, a_2, q_2) \in F_2\} \end{aligned}$$

Besides the treatment of the end places, the above construction is similar to [7, 17]. In [17], however, the root-unwinding step introduces $2^{\#P_{cyc}}$ new transitions, with P_{cyc} the set of initial places that are in cycles. In [7], an extra restriction applies: all transitions that have initially marked input places must be enabled.

Parallel composition. In Petri nets a transition can be regarded as a synchronization mechanism since it can only fire if all input places contain at least one token. To model parallel composition with rendezvous synchronization, it is then sufficient to "join" the transitions that have the same action label, different from τ. Since more than one transition may be labeled with the same action, all combinations have to be considered. This is shown in Figure 4.

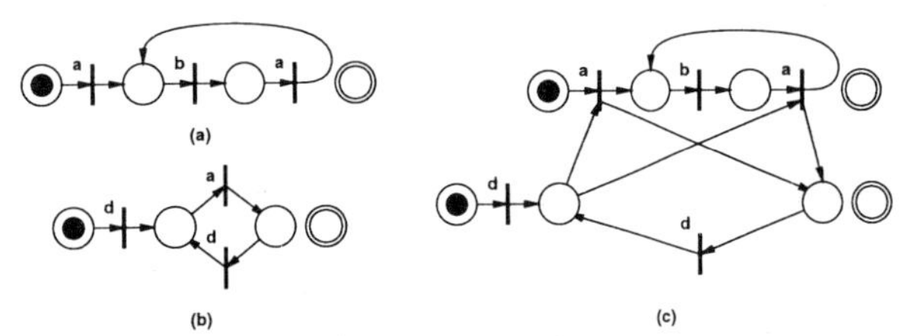

Figure 4: Petri net representations (a) $\mathcal{PT}(Q_1)$ (b) $PT(Q_2)$ (c) $PT(Q_1 || Q_2)$

Definition 3.10 (Parallel Composition) *Let Q_1 and Q_2 be two processes, with $\mathcal{PT}(Q_1) = \langle P_1, A_1, F_1, M_{0_1}, E_1\rangle$ and $\mathcal{PT}(Q_2) = \langle P_2, A_2, F_2, M_{0_2}, E_2\rangle$. $\mathcal{PT}(Q_1 || Q_2)$ is defined as:*

$$\langle P_1 \cup P_2, A_1 \cup A_2, F', M_{01} \cup M_{02}, E_1 \cup E_2\rangle$$

where

$$\begin{aligned} F' = \ & \{(I, a, O) \in F_1 \cup F_2 | a \notin (A_1 \cap A_2) \setminus \{\tau\}\} \\ & \cup \{(I_1 \cup I_2, a, O_1 \cup O_2) | \ a \in (A_1 \cap A_2) \setminus \{\tau\} \\ & \quad \wedge (I_i, a, O_i) \in F_i\} \end{aligned}$$

Besides the treatment of τ events, the above construction is similar to [7].

Interrupt Composition. For the PE-net representation of $Q_1 \rhd Q_2$, we create a new place for every transition of $\mathcal{PT}(Q_1)$. Each new place is then connected via a self-loop with its corresponding transition. These new places are then combined with the initial places as well as the end places of $\mathcal{PT}(Q_2)$, by means of a Cartesian product construction. If Q_2 is a recursive process expression, the one-step unfolding within $\mathcal{PT}(Q_2)$ (see Definition 3.12) then prevents the transitions of $\mathcal{PT}(Q_1)$ from firing for every loop iteration within the former net. However, if Q_2 is of the form $(P \rhd R)\ldots$, within $\mathcal{PT}(Q_2)$ there can still be initial places that are in cycles (see Definition below). In this case an extra root-unwinding step is necessary. For the other cases, this preprocessing step has no effect. This construction is shown in Figure 5.

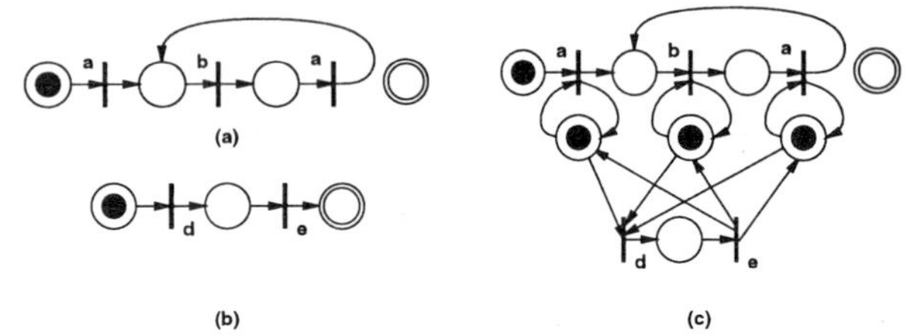

Figure 5: Petri net representations (a) $\mathcal{PT}(Q_1)$ (b) $\mathcal{PT}(Q_2) = RootUnw(\mathcal{PT}(Q_2))$ (c) $\mathcal{PT}(Q_1 \rhd Q_2)$

Definition 3.11 (Interrupt Composition) *Let Q_1 and Q_2 be two processes, with $\mathcal{PT}(Q_1) = \langle P_1, A_1, F_1, M_{0_1}, E_1\rangle$ and $RootUnw(\mathcal{PT}(Q_2)) = \langle P_2, A_2, F_2, M_{0_2}, E_2\rangle$. Let P_{new} be new places such that $P_1 \cap P_2 \cap P_{new} = \emptyset$, and H a bijection $\mathrm{H} : F_1 \to P_{new}$. $\mathcal{PT}(Q_1 \rhd Q_2)$ is defined as:*

$$\langle P_1 \cup (P_2 \setminus (M_{02} \cup E_2)) \cup (P_{new} \times M_{02} \times E_2), A_1 \cup A_2, F', M_{0_1} \cup (P_{new} \times M_{02} \times E_2), E_1\rangle$$

where

$$\begin{aligned} F' = \ & \{(p_1 \cup p_n, a_1, q_1 \cup p_n) \mid (p_1, a_1, q_1) \in F_1 \\ & \quad \wedge p_n = \mathrm{H}(p_1, a_1, q_1) \times M_{02} \times E_2\} \\ & \cup \{(p_2 \setminus M_{02} \cup (P_{new} \times (p_2 \cap M_{02}) \times E_2), a_2, \\ & \quad q_2 \setminus E_2 \cup (P_{new} \times M_{02} \times (q_2 \cap E_2))) \mid \\ & \quad (p_2, a_2, q_2) \in F_2\} \end{aligned}$$

Recursion. For the PE-net representation of a process $\mu x.F(x)$, with $F(x)$ a process expression containing variable x, we first compute $\mathcal{PT}(F(x))$, using the translation techniques described above. The input places of each transition with label x are then combined with the initial places, by means of a Cartesian product, effectively creating the desired loop. The initial places as well as the initial marking are kept; this construction implements a one-step unfolding making the root-unwinding step obsolete for the PE-net translation of process expressions of the forms $P + P$ and $Q \rhd P$, with P a recursive process definition. This is shown in Figure 6.

Definition 3.12 (Recursion) *Let $F(x)$ be process expression with process variable x, constructed by means of sequential composition, (non-)deterministic choice composition, parallel composition and interrupt composition, and*

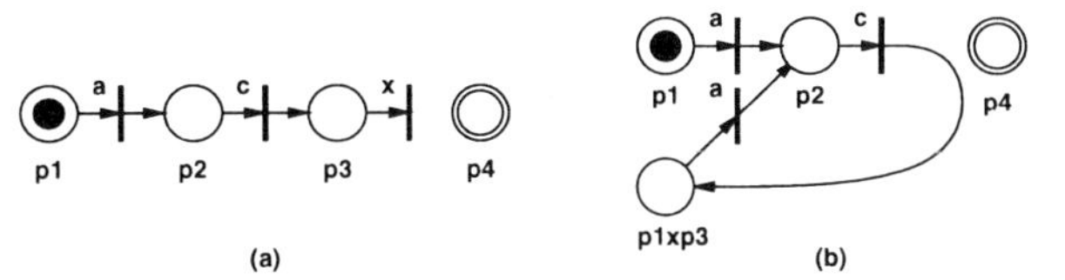

Figure 6: Petri net representations (a) $\mathcal{PT}\ (F(x))$ (b) $\mathcal{PT}\ (\mu x.F(x))$

$\mathcal{PT}\ (F(x)) = \langle P, A, F, M_0, E\rangle$. $\mathcal{PT}\ (\mu x.F(x))$ *is defined as:*

$$\langle (P \setminus P_x) \cup (M_0 \times P_x), A \setminus \{x\}, F', M_0, E\rangle$$

where

$$P_x = \bigcup_{(p,x,\emptyset)\in F} p$$

$$\begin{aligned} F' = \ & \{(p \setminus M_0 \cup (p \cap M_0) \times p_x, a, \\ & \quad q \setminus P_x \cup M_0 \times (q \cap P_x)) \mid a \neq x \\ & \quad \wedge (p,a,q) \in F \wedge (p_x, x, \emptyset) \in F\} \\ & \cup \{(p, a, q \setminus P_x \cup M_0 \times (q \cap P_x)) \mid a \neq x \\ & \quad \wedge (p,a,q) \in F\} \end{aligned}$$

3.3. Translation of Process Calculus expressions

The translation of a (complex) process expression into a Petri net can be defined recursively as follows. We start at the bottom of the syntax tree by translating the "leaf" atomic actions to PE-nets, according to definition 3.5. As we go up in the syntax tree, we gradually build up the PE-net representations of the intermediate subexpressions. This is illustrated in the example of Figure 7. Consider

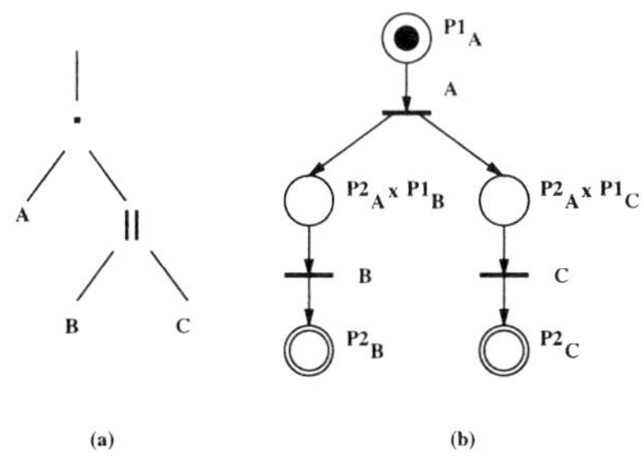

Figure 7: (a) Syntax-tree of $A.(B||C)$ (b) $\mathcal{PT}\ (A.(B||C))$

the process expression $A.(B||C)$. This expression has the simple syntax tree shown in Figure 7(a). Starting at the bottom of the tree, the leaf actions are A, B and C with corresponding PE-nets $\mathcal{PT}\ (i) = \{\{p1_i, p2_i\}, \{i\}, \{(\{p1_i\}, i, \{p2_i\})\}, \{p1_i\}, \{p2_i\}\}$ for $i \in \{A, B, C\}$. As we go up in the syntax tree, we first encounter the parallel composition operator, and we can compute $\mathcal{PT}\ (B.C)$ as an intermediate result. Next, as we go up, we arrive at the sequential composition operator, and use the intermediate result to compute $\mathcal{PT}\ (A.(B||C)$, which is depicted in Figure 7(b).

4. Verification Framework

The techniques presented in this paper have been implemented in a tool, called JULIE, which consists of about 9000 lines of C code. The tool starts from the process calculus, applies a syntax-directed mapping scheme to translate between process expressions and Petri nets and performs an efficient analysis on the resulting Petri nets (e.g. checks for deadlocks, liveness properties, etc). Currently, we are developing an automated translation between heterogeneous C-VHDL specifications, as they are used in the CoWare^{TM}data model, and our process calculus. Because the overall mapping is semantics preserving, the analysis technique can be applied to validate the various intermediate specifications of the design process.

The analysis technique itself, called *generalized partial-order analysis* [18], tackles the two primary sources of combinatorial explosion that may occur in conventional reachability analysis. The first source is due to concurrently enabled actions for which standard analysis requires enumerating all possible orderings. This problem can be avoided by e.g. applying existing partial-order techniques where only one interleaved sequence needs to be analyzed for deadlock and liveness checks [15]. The second source is due to concurrently marked conflict places. This problem is solved by a generalized partial-order method which explores simultaneously concurrently enabled conflicting paths. The technique is based on a modified representation of markings to distinguish the different conflicting paths and can achieve an exponential reduction in algorithmic complexity.

5. Case study

To assess the viability of our approach we performed an extensive case study. More specifically, we experimented with the design of a Quadrature Amplitude Modulation (QAM) modem, integrating both a sender and a receiver section. The block diagram of the modem design is depicted in Figure 8. In the sender part (bottom square) the data to be sent is first formatted into a stream of alternating *I-data* and *Q-data*, each three bits wide. Then, this data is discretely levelized by the slicer and sent to the modulation block, which multiplies the I-data and the Q-data by orthogonal carriers. The middle square of Figure 8 represents the test bench including the channel model and the user interface. The upper square represents the receiver part. A tracking module is needed to derive "a local copy" of the carrier frequency, as well as an n-taps adaptive equalizer for correcting channel induced distortion. The de-slicer and symbol extraction block perform the inverse operations of their sender counterparts.

The QAM modem was modeled using the CoWare^{TM}environment [3]. The implementation target was a heterogeneous single-chip solution; because the (de)slicer and symbol creation (extraction) modules are dependent on

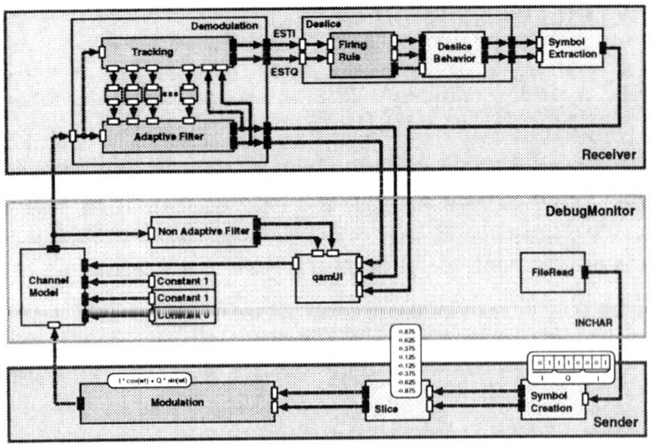

Figure 8: Block diagram of the QAM modem

the characteristics of the data, these modules were chosen to be implemented in software on an ARM-7 RISC processor. The other modules were specified in VHDL to be implemented in application-specific hardware. Simulations and formal verification using the techniques presented in the paper, revealed that the heterogeneous C-VHDL specification was indeed deadlock free.

To implement the communication between the hardware and the software, the ARM-7 processor boundary has to be crossed. This is shown in Figure 9 for the communication between the tracking module of the demodulator and the de-slicer. To realize the software drivers and the hardware interface, as depicted in Figure 9(b), the Symphony toolbox[3] has a number of I/O scenarios to select from. An I/O scenario (e.g. memory mapped I/O, interrupt driven I/O) consists of a software driver and a hardware counterpart for implementing a specific channel type on a particular processor. The Symphony toolbox assigns I/O scenarios on a per-channel basis. The I/O scenarios are designed to work correctly in a stand-alone operation mode. However, when assembled to a complete hardware/software interface, deadlocks may be introduced for a particular combination of I/O scenarios. Exhaustively verifying all these combinations by simulation is very time consuming as a new compilation step is required for every investigated I/O-configuration. Selecting the right I/O-configuration simply by inspection is quasi impossible, as the "real" problem instance is far more complex than what is presented here. Our verification approach formally diagnoses the cause of a possible deadlock, and provides the necessary feedback to select the right I/O scenario combination(s). To get an insight in what this means, an example is worked out below.

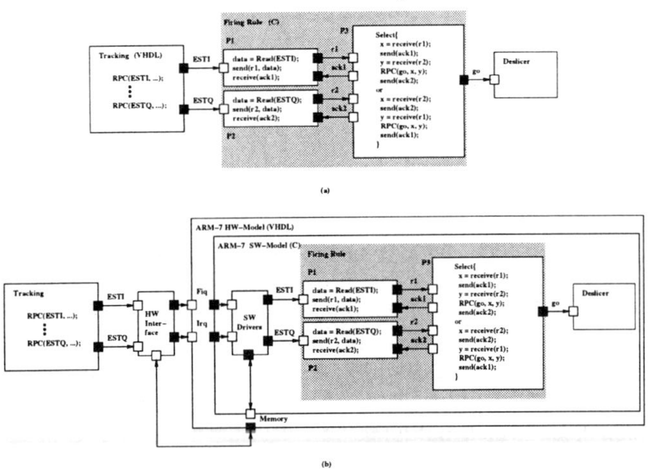

Figure 9: (a) Simplified view of communication between tracking module and de-slicer; (b) Implementation scheme of communication

Consider the following I/O-configuration. The symbol creator of the sender part receives incoming samples from the INCHAR channel via software polling. Processes $P1$ and $P2$ of the de-slicer (see Figure 9(a)), on the contrary, are assigned to the FIQ and IRQ interrupt routines of the ARM-7, respectively. The corresponding hardware interface does not wait for the interrupt routines to complete; they are only "triggered".

Without loss of generality, we focus on the communication between the tracking module and the de-slicer. This can be modeled in our process calculus as follows:

```
/* Software */
P1 = r1.ack1;
P2 = r2.ack2;
P3 = ((r1.ack1.r2.Deslice.ack2) +
      (r2.ack2.r1.Deslice.ack1)).P3;
SOFTWARE = P1 || P2 || P3;
/* Software drivers */
IRQ  = P1; /* P1 -> IRQ routine */
FIQ  = P2; /* P2 -> FIQ routine */
/* Main Body */
MAIN = ; /* Empty ! */
/* Processor Model of ARM-7 */
ARM = (MAIN > (IRQ_B.IRQ.IRQ_E))
                       > (FIQ_B.FIQ.FIQ_E);
/* Hardware Interface */
EST_I = FIQ_B; /* I/O-scenario fiq_trigger */
EST_Q = IRQ_B; /* I/O-scenario irq_trigger */
/* Tracking block of Demodulation */
TRACKING = EST_I.EST_Q.TRACKING;
SYSTEM = SOFTWARE || ARM || TRACKING;
```

The suffixes _B and _E denote the triggering and the completion of the interrupts involved. `FIQ` and `IRQ` represent the (calling of) the respective interrupt routines themselves. The (simplified) processor model of the ARM-7 models that the FIQ-interrupt has a higher priority than the IRQ-interrupt; in other words, the former interrupt can preempt the latter. If we run this example through JULIE we get the following reachability graph.

```
> Reach(SYSTEM);
Deadlock in State9   !
```

[3] Symphony is part of CoWare™ and is responsible for the hardware/-software interfacing and system architecture co-synthesis and integration problems.

```
STATE TRANSITIONS:
{
( State0        { FIQ_B }       State1 )
( State1        { r2 }          State2 )
( State2        { ack2 }        State3 )
( State3        { FIQ_E }       State4 )
( State4        { IRQ_B }       State5 )
( State5        { r1 }          State6 )
( State6        { Deslice }     State7 )
( State7        { ack1 }        State8 )
( State8        { IRQ_E }       State0 )
( State5        { FIQ_B }       State9 )
}
```

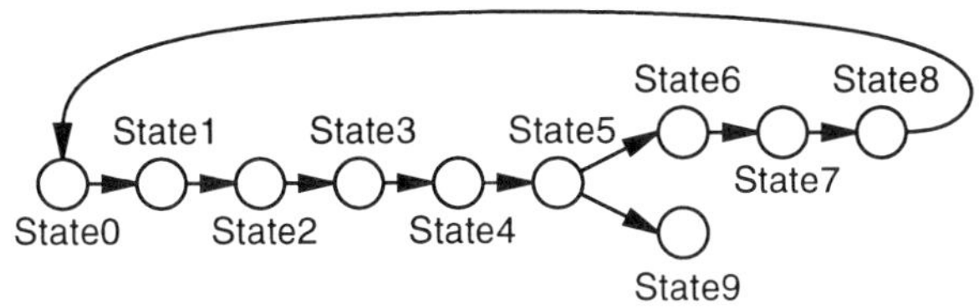

From state `State9` in the table above, no progress is observed which shows that this particular I/O configuration results in a deadlock. Indeed, after the first FIQ-routine has finished, the IRQ-routine is called. However, this latter routine, before being effective, can be preempted immediately by a new FIQ interrupt, which has a higher priority, and as result we arrive in a deadlock situation.

In Table 1, a number of possible I/O configurations are listed. As can be observed, verifying a particular configuration takes less than 1 second CPU time. From the table one can conclude that two configurations can result in a deadlock. For selecting the "optimal" I/O configuration, Symphony only pursues with the remaining alternatives. This result clearly shows that the process calculus and the subsequent Petri net translation, being able to model all communicating processes involved as well as the processor model of the ARM-7 itself, is indeed a powerful combination for analysis and synthesis of embedded systems.

Table 1: I/O configurations for implementing the hardware interface and the software drivers of Figure 9(b)

I/O scenarios			dead-	time
INCHAR	ESTI	ESTQ	lock	(sec)[4]
sw-polling	irq-trigger	fiq-trigger	NO	0.46
sw-polling	fiq-trigger	irq-trigger	YES	0.41
sw-polling	irq-wait	fiq-trigger	NO	0.36
sw-polling	fiq-trigger	irq-wait	NO	0.35
sw-polling	irq-trigger	fiq-wait	NO	0.44
sw-polling	fiq-wait	irq-trigger	YES	0.36
sw-polling	fiq-wait	irq-wait	NO	0.33
sw-polling	irq-wait	fiq-wait	NO	0.37

[4] Run-times measured on a HP 900/715 64 MB.

6. Conclusions

In this paper we proposed a formal model based on Petri nets for reasoning about the behavior of a concurrent system. Using this Petri net level, existing formal verification techniques can then be leveraged.

To provide a formal mapping between language-level constructs and Petri net operations, we proposed a process calculus as an intermediate model. We believe many of the existing process-based languages, used to specify heterogeneous systems can be first translated to "intermediate code" of our process calculus. We presented an elegant syntax-directed translation scheme for building Petri net representations starting from process expressions.

The viability of our approach was tested, on a "real-life" example, for which the results were very promising. We are currently investigating the inclusion of explicit timing constraints. In the future we plan to fully integrate the presented verification framework into the Symphony design flow.

References

[1] J.A. Bergstra and J.W. Klop. Algebra of Communication Process with Abstraction. *Theoretical Computer Science*, 37(1):77–121, 1985.

[2] I. Bolsens, H. De Man, B. Lin, K. Van Rompaey, S. Vercauteren, and D. Verkest. Hardware/software Co-Design of Digital Telecommunication Systems. *Proceedings of the IEEE*, 85(3):391–418, March 1997.

[3] CoWare, 2385 Santa Ana Street, Palo Alto, CA 94303, USA and Kapeldreef 60, B-3001 Heverlee

[4] L. Philips I. Bolsens and H. De Man. A Programmable CDMA IF Transceiver ASIC for Wireless Communications. In *IEEE Custom Integrated Circuits Conference*, May 1995.

[5] J. R. Burch, E. M. Clarke, K. L. McMillan, and D. L. Dill. Sequential Circuit Verification using Symbolic Model Checking. In *Proceeding of the 27th ACM/IEEE Design Automation Conference*, Orlando, June 1990.

[6] O. Coudert, J. C. Madre, and C. Berthet. Verifying Temporal Properties of Sequential Machines without building their State Diagrams. In *Workshop on Computer-Aided Verification*, Rutgers, June 1990.

[7] G. G. de Jong and Bill Lin. A Communicating Petri Net Model for the Design of Concurrent Asynchronous Modules. In *Proceedings of the 31rd ACM/IEEE Design Automation Conference*, June 1994.

[8] B. Gyselinckx, L. Rijnders, M. Engels, and I. Bolsens. A 4*2.5Mchip/s Direct Sequence Spread Spectrum Receiver ASIC with Digital IF and Integrated ARM6 Core. In *IEEE Custom Integrated Circuits Conference*, May 1997.

[9] C. A. R. Hoare. *Communicating Sequential Processes*. Prentice-Hall, 1985.

[10] R. Milner. *Communication and Concurrency*. Englewood Cliffs, NJ, Prentice Hall, 1981.

[11] International Standards Organization. *LOTOS: a Formal Description Technique based on the Temporal Ordering of Observational Behavior*. ISO, IS 8807, 1989.

[12] E. Pastor, O. Roig, J. Cortadella, and R. M. Badia. Petri Net Analysis using Boolean Manipulation. *Lecture Notes in Computer Science 815, Springer Verslag*, pages 416–435, June 1994.

[13] J.L. Peterson. *Petri Net Theory and Modelling of Systems*. Prentice Hall, 1981.

[14] D. Taubner. Finite Representations of CCS and TCSP Programs by Automata and Petri Nets. *Lecture Notes in Computer Science 369*, June 1989.

[15] A. Valmari. A Stubborn Attack on State Explosion. In 2nd *Workshop on Computer Aided Verification*, pages 156–165, New Brunswick, NJ, June 18-21 1990.

[16] K. van Berkel. *Handshake Circuits: An Asynchronous Architecture for VLSI Programming*. Cambridge International Series on Parallel Computation, 1993.

[17] R. van Glabbeek and F. Vaandrager. Petri Net Models for Algebraic Theories of Concurrency. *Proc. of PARLE II, Lecture Notes in Computer Science 259*, pages 224–242, 1987.

[18] S. Vercauteren. Efficient Detection of Deadlocks Using Generalized Partial Order Analysis. *IMEC Internal Technical Report, Reference: IMEC/VSDM/97Mar/Deadlock/gpo.ps*, April 1997.

[19] S. Vercauteren and Bill Lin. Hardware/software Communication and System Integration for Embedded Architectures. *Design Automation of Embedded Systems, Kluwer Academic Publishers*, 2:1–24, 1997.

Session 2

Fast Prototyping and Code Generation

Prototyping of the Receiver Unit for a Broadband Access Network

A. Hein, J. Dalcolmo, P. Le Corre, R. Lauwereins, M. Adé
Kard. Mercierlaan 94
K.U. Leuven
ESAT/ACCA Laboratory
B-3001 Heverlee, Belgium
email: Axel.Hein@esat.kuleuven.ac.be

Abstract

To exploit the potential of two-way communication on CATV networks for advanced interactive telecommunication applications like video-on-demand, a high speed modem for up-stream communication is currently under development. We are using the rapid prototyping tool GRAPE to evaluate the digital receiver part of the modem which is described in detail in this paper. The sampling rates achieved for a 16-QAM modem on multiple DSP processors are reported.

1. Introduction

Recently, coaxial cable networks have received much attention in the context of interactive applications [3, 5]. In countries with a high penetration of CATV (in Belgium more than 90%), it may serve as an alternative to classical telephone networks. The envisioned applications are telephony, interactive television, home-shopping, video-on-demand, high-speed Web browsing, etc [7]. Since these applications require the possibility of interactive use, two-way communication has to be provided on the networks which initially have been designed for only a one-way broadcasting of television signals.

What is aimed at now, is a 10 Mbit/s upstream link (from the subscriber to the head-end station) and a downstream link (from the head-end station to the local subscribers) with higher bit rate. Especially the upstream is challenging because very little is known about the upstream channel, and communication standards are still under development [4]. The projected frequency band for upstream communication is in the 5-25 MHz range.

Since the basic configuration of the network is hybrid fibre-coax, the upstream signal will first travel through coax before entering a fibre node and going through the fibre trunk to the head-end station. The coaxial part will bring along some serious channel impairments which have to be compensated for at the receiver. The most important impairments are mentioned below without going into too much detail. Group delay distortion (ie signals at different frequencies propagating with a different velocity) causes severe inter-symbol interference at the receiver. Micro-reflections are caused by discontinuities in the transmission medium and cause part of the signal energy to be reflected. Ingress noise models the interference caused by the antenna-like properties of the cable. Burst noise typically originates from household appliances such as electrical motors. Besides this there are common path distortion products, thermal noise, impulse noise, non-linearities, phase-noise frequency offset etc. Adding to this the variations in the network stemming from the variability of the number of trunk, bridge and distribution amplifiers it becomes quite clear that it is very hard to build a channel model which incorporates all these statistical and non-statistical phenomena observed in real networks. Studies [5] have shown that due to this inhospitable environment, less than half of the spectrum will be available at any given time instant. The absence of a good mathematical channel model necessitates the *real-time prototyping* of the complete set-up, including the cable, after off-line simulation and before commitment to silicon is made. Real-time prototyping has several benefits compared to off-line simulation:

- It enables algorithm verification on the real channel.
- It allows for more extensive testing under more varying conditions: 24 hours a day measurements

1080-1820/97 $10.00 © 1997 IEEE

are possible as well as tests with transmitters placed on different locations.

- It gives faster feedback when modifying algorithmic settings: the prototype allows to modify algorithmic parameters on the fly, without recompilation and to view its effects in real-time on the next received data packet.

We aim at developing a 16-QAM modem for the upstream communication channel described above. The projected bit rate is 10 Mbit/s at an overall bit-error rate of 10^{-10} including software-based error correction. The transmission payload consists of ATM cells and the multiple access protocol is TDMA.

A prototype of the receiver part of a 16-QAM modem is currently under development at the ESAT/ACCA research group of the K.U. Leuven. A sketch of the modem architecture is shown in Figure 1. The algorithm for the decoding of the data has been simulated and is to be implemented in an ASIC. It is desirable to test the algorithm on data on a real network, and to measure bit and frame error rates. However, workstation simulation of the algorithm is in the range of 10^4 slower than real time operation and therefore too slow for such an evaluation. It is rather obvious, that fast execution time of the prototype is mandatory; this can be done by porting the algorithm to a set of DSP processors or FPGAs. Since DSPs provide larger flexibility, the algorithm of the receiving unit has been ported to a set of DSPs, working in parallel, in order to achieve suitable performance. GRAPE has been chosen as the rapid prototyping tool to implement the algorithm on a set of DSPs, because it allows automatic task assignment, routing, and scheduling, gives rapid estimation of the performance, and automatically generates the code required for the DSPs [6].

Since the software running on a set of DSPs is still not capable to provide real time data throughput, the problem had to be divided into a software and hardware domain. The software runs the actual receiver algorithm which is to be tested, the hardware provides the access to real time sampling rates, at least for the duration of one or a few data bursts, ie data for a single network user can be processed in real time.

In the following sections we present the components of the receiver unit of the modem – the Sample Rate Reducer (section 2) and the processing unit of the receiver (section 3). Measurements are presented in section 4, and the main aspects of the paper are summarized in the final section 5.

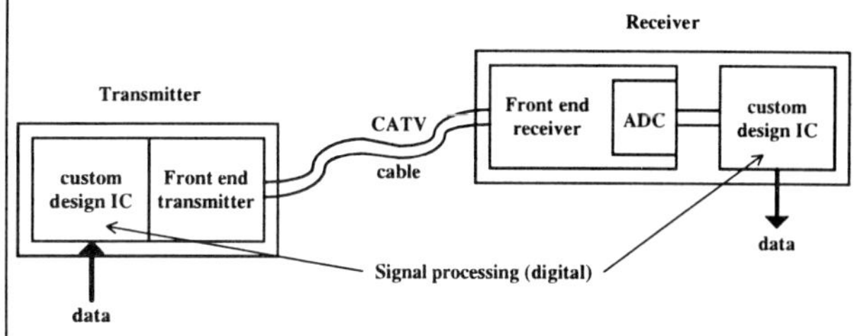

Figure 1. Modem Setup

2. Sample Rate Reducer

The modem algorithm running on a set of DSPs cannot keep up with the rate of incoming data in real time. However, in order to test the algorithm on a real network, real time samples have to be captured and played back slower to the DSPs running the algorithm under test. This is the task of the Sample Rate Reducer shown in Figure 2. This device, developed by the ESAT/ACCA group at K.U. Leuven, captures a frame of real time data, stores it, and plays it back to the DSP boards. The number of samples per frame is programmable, and the Sample Rate Reducer has features to facilitate debugging.

Furthermore, the hardware has the task of synchronization with the data bursts: The energy level on the net is calculated and a discriminator triggers on the beginning of a data burst. Since this must happen in real time, it could not be implemented in software. Synchronization is needed to avoid storage and software processing of long periods of samples between data bursts. Elimination of these periods is what actually provides the reduction of the sample rate needed for software emulation of the receiver algorithm.

2.1. Sample Rate Reducer Concept

The main function of the Sample Rate Reducer is that of a frame grabber. In order to test the modem receiver algorithm on the DSPs, samples with predefined content are sent by a test transmitter. The data arrives from the receiver ADC at a sampling rate of approximately 10 MHz and is then stored in two circular buffers simultaneously. One of these is used to buffer data flowing to the DSPs, the other one is buffering a

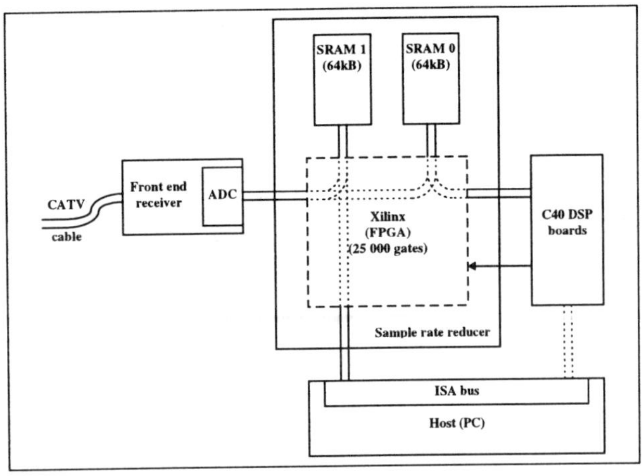

Figure 2. Sample Rate Reducer

copy for debugging purposes and allows downloading of the incoming data to the host PC simultaneously with new data acquisition. When a frame of data has been detected, the data acquisition is temporarily suspended, and one copy of the data, including pre-trigger data is transferred to the DSPs. The DSPs run the modem algorithm on it, then compare it with the expected data content and return at the completion a flag indicating success or failure of the reception.

If the flag returned by the DSPs indicates an error, the second copy of the data is moved to the host computer via an ISA bus interface, so it can be processed off line by feeding it back into simulation to improve the receiver algorithm.

Thus, the better observability of simulation models is combined with the higher speed and 24-hours-a-day unattended measurement capabilities of prototyping.

Even while a frame is being downloaded to the host, the Sample Rate Reducer restarts with collection of new data, now only saving one copy of incoming data, until the host download is completed. In this fashion, there is no interruption in the data flow, even when debugging information is being dumped to the host computer. Naturally it is not possible in this way, to provide debugging information on a continuous basis, but only for the first of a set of erroneously received frames.

The net result is that the output data rate is in the order of about 200 times slower than the incoming rate, when using a single DSP, which means that one out of 200 frames could be analyzed, if the incoming data stream would be near continuous. The actual rate is dependent on the number of DSPs used to run the receiver algorithm. For test purposes, the repetition rate of the frames on the network is reduced by a factor of 200, but the frames themselves are sent with real time speed.

2.2. Implementation Details

Unfortunately it was not possible to use the rapid prototyping environment GRAPE to implement the data flow through the FPGA, as has been done for the DSP part, because the necessary VHDL code generator within GRAPE is still under development. Therefore we had to resort to code the FPGA by hand, using VHDL.

All the logic on the board of the Sample Rate Reducer, except for an ISA Bus address decoder, is implemented in a Xilinx FPGA chip. All the logic on this chip was designed in VHDL.

Besides the 16 bit ISA bus interface, there are TMS320C40 style DSP links implemented, to communicate with the DSPs, and a general purpose parallel interface to capture modem data coming from an ADC. All I/O interfaces are electrically buffered. The FPGA has direct simultaneous access to two 16 bit wide, 64 kWord deep static RAM chips, with 15ns cycle time.

The data flow is from the parallel interface through the FPGA to the two RAMs, then from one RAM via the FPGA to a link to the DSP boards, and for the debugging information from the second RAM via the FPGA to the ISA bus. Various parameters like the frame length and repetition rate are programmable via the ISA bus interface.

2.3. FPGA

The FPGA is in circuit re-programmable via a standard Xilinx XChecker interface, and can also be probed via this interface. In addition there are a few facilities on board to ease debugging of the board or any design inside the FPGA.

The design programmed into the FPGA is modular and consists of several blocks shown in Figure 3.

- **Energy Calculation and Burst Detection**: The signal energy on the network cable can be calculated by $I^2 + Q^2$, where I and Q are a pair of successive ADC samples. The energy calculation is implemented as an approximation to avoid the high hardware cost of multipliers. The energy calculated from the pairs of samples is averaged over four samples and compared with a threshold. If this average is above theshold four times consecutively, it is assumed to be the beginning of a data burst. Since the 12-bit wide samples arrive at a

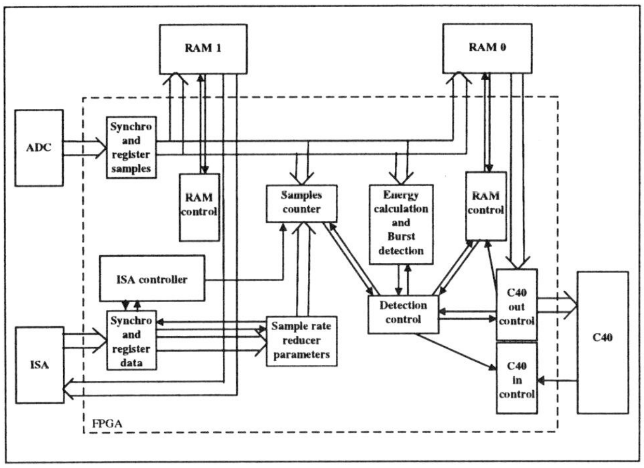

Figure 3. Overview of the Receiver Component

rate of 10 MHz, the speed of the energy calculation has to be taken into account too. The various operations that take place are pipelined in order to achieve the required throughput.

- **ISA Controller**: A PC interface to a set of parameter registers and to read status flags as well as captured data frames for debugging.

- **Sample Rate Reducer Parameters**: A set of registers for: the threshold value for burst detection, the size of the data frame, the trigger position, the time to wait for the DSPs etc.

- **Sample Counter**: A count down counter decrementing at each incoming sample advances the state machine when it reaches 0. It is used to position the trigger within the data, to stop and start data aqcuisition and to time the waiting period for the DSPs.

- **Detection Control**: This is the top level statemachine of the Sample Rate Reducer, and driven mainly by the state of the Sample Counter, the input from the Burst Detection and the DSPs.

- **RAM Controllers**: Essentially silicon tape recorders, storing the incoming data at high speed to replay them slowly.

- **C40 Controllers**: The DSP word length is 32 bits which the link sends asynchronously in four times 8 bits. Both blocks handle the hand-shaking signals and convert the data format (the FPGA uses a 12-bits two's complement representation of the data).

- **Synchro and Register**: The ISA bus is running asynchronously with the FPGA, and there is an unknown difference in phase between the ADC and the FPGA. These blocks are used for synchronization.

3. Processing Unit of the Receiver

The processing unit of the receiver is responsible for the decoding of the incoming symbols and for the post-processing which includes checking for bit errors and notifying the sample rate reducer in the case of errors.

The tasks of the processing unit as well as their interdependencies are defined within the GRAPE environment as an extended data-flow model which is called *cyclo-static data-flow* (CSDF) [2] and represents an extension of Lee's *synchronous data-flow* (SDF) [8]. CSDF maintains all the properties of SDF but allows for the consideration of *cyclically changing behavior* such as time-multiplexed channels with a static and cyclically repeated period. Although this cyclo-static behavior is known at compile time and is independent of any run-time data, it cannot be specified within the SDF framework which requires fixed and constant token productions and consumptions.

GRAPE's design flow consists of several phases [1] as shown in Figure 4. Regarding the data-flow representation, the application has to be set up as a directed graph $G=(N, E)$, where the nodes N represent computation tasks, and the edges E depict the communication of the results (commonly called the *tokens*) between a producing and a consuming task. The functionality of the nodes is specified in a conventional high-level language like C (for DSP processors) or VHDL (for FPGAs). The number of tokens a task consumes and produces during an execution phase of a task is known at compile time, allowing for a compile-time analysis of the graph in the next phases of GRAPE's design flow and leading to highly efficient run-time code. GRAPE embeds the task descriptions in shells which handle all the communication.

The target hardware is represented as a connectivity graph with an indication of the amount and type of resources each processing device possesses. In the following step of GRAPE's design flow, the amount of resources required by each of the tasks when executed on each of the processing devices, is estimated. Then, the application is mapped onto the target hardware, ie the assignment, routing, and scheduling of the tasks are automatically optimized to minimize the buffer lengths and the total execution time. The final code is generated depending on the target hardware, an estimation of the execution times is given, and the application can be downloaded on the target hardware. It should be emphasized that the target hardware can be changed rapidly without modifying the data-flow model of the application and vice-versa. This independence of application and target hardware has tremendously facilitated the performance measurements of the application on different hardware platforms as will be presented in section 4.

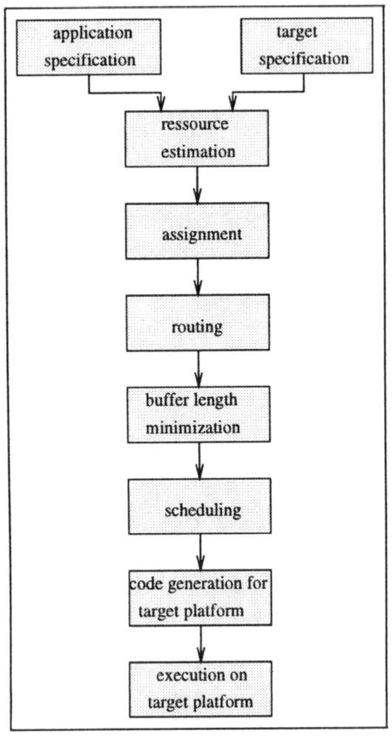

Figure 4. Flow of GRAPE Design Process

The architecture of the 16-QAM receiver unit which has been designed by IMEC and the ESAT/SISTA research group consists of the following tasks:

- **DMUX** (Demultiplexer) : The 4-phase demultiplexer scales the incoming values before they are forwarded.
- **FIL[IQ]** (FIR Filter) : These two FIR filters downsample and filter the incoming data by performing a convolution with pre-computed parameters.
- **AGC** (Automatic Gain Control) : The AGC controls and corrects the gain. The changes between the internal idle and active states depend on the input data and state parameters.
- **EQ** (Burst Acquisition Equalizer) : The EQ task performs the detection of symbol sequences. The state changes depend on the input data and state parameters.
- **LMS** (Symbol-Spaced Adaptive Equalizer) : Equalizing is performed based on feedforward and feedback convolutions.
- **DEMAP** (Demapper) : The constellations of 16-QAM or 4-QPSK, respectively, are mapped back to symbols. The 4-QPSK encoding is provided as an alternative to 16-QAM but is not further taken into account within this paper.

Figure 5 gives an overview of the receiver unit represented as a data-flow model.

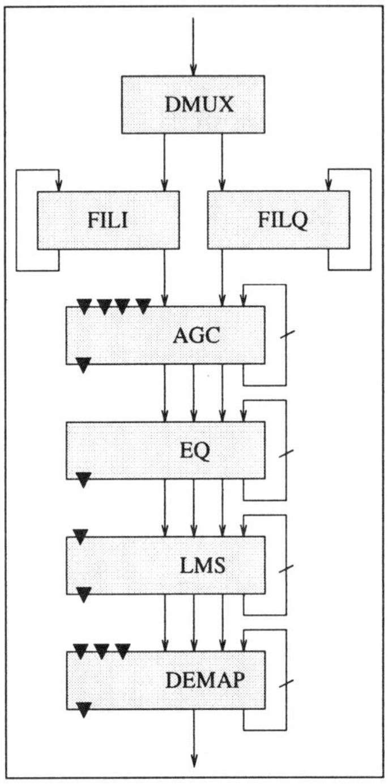

Figure 5. Data-Flow Model of the Application

During each static schedule DMUX is executed four times, and all the other tasks are triggered twice, ie the incoming data are downsampled. In each of the four executions of DMUX another set of operations is started. The order of these four phases is known at compile-time and can be statically assigned and scheduled; the consideration of this cyclo-static behavior (CSDF) is one of the outstanding characteristics of the GRAPE tool.

The execution of LMS and DEMAP is data-dependent, ie they are triggered by EQ. LMS and DEMAP perform operations only when *useful* data is coming in, and in that case only once per schedule; therefore, EQ downsamples by a factor of two. A static assignment tool as it is used by GRAPE cannot handle this data-dependent execution which can be considered as dynamic data-flow. Hence, LMS and DEMAP are triggered twice a schedule, such as AGC and EQ, but the incoming data provides information whether operations have to be performed at all.

4. Performance Measurements

The original code of the application has been optimized to speed up the execution times on the DSP processors. This step clearly increased the sampling rate of the application. Table 1 shows the sampling rates for simulation on a SUN UltraSparc Workstation and for one, two and four DSP processors TMS320C40 before (T_{no-opt}) and after (T_{opt}) code optimization, which includes elimination of redundant operations, loop unrolling, and introduction of additional variables to avoid address calculations. Moreover, expensive `switch` conditions have been replaced by nested `if` statements.

Architecture	T_{no-opt} [kHz]	T_{opt} [kHz]
Simulation	0.19	–
1 C40	12.3	22.7
2 C40	23.6	42.4
4 C40	23.6	70.4

Table 1. Sampling Rates of the Processing Unit

When moving from 2 DSPs to 4 DSPs for the unoptimized algorithm, no improvement of the execution time can be achieved. This is due to the fact that the AGC task takes about the same execution time per schedule as all the other tasks together; in its original version, the shifting of numerous state parameters is performed by copying them within arrays. By replacing this expensive mechanism with a few additional variables pointing to the *logical* begin of the arrays, the copy operations can be avoided and the execution time of the AGC was reduced from 1624 down to 356 cycles.

Several *runtime-* and *back-parameters* have been added to the application. Runtime-parameters make it possible to modify algorithmic parameters at runtime, whereas back-parameters provide instantaneous feedback from the application running on the DSPs to the user or a control program running on the PC (in Figure 5 the runtime-parameters are drawn as black triangles pointing to the corresponding block, and the back-parameters are represented as dark-grey triangles pointing out of the block). The automatic gain control (AGC) can be bypassed to set the gain value manually, and threshold values may be modified at runtime. Also, the modulation type (4-QPSK or 16-QAM) can be set. Additionally, a derandomizer can be activated and initialized (DEMAP). The implemented back-parameters provide information about the gain value (AGC), the tap coefficients (EQ and LMS), and the correctness or error of the received and decoded samples (DEMAP).

The runtime- and back-parameters allow for testing and monitoring of the application. When problems are encountered, these values are used to improve the application by going back to more flexible but slower simulation runs.

5. Summary

Although continuous real-time sampling rates cannot be achieved, prototyping is shown to be a needed and useful means to evaluate the interaction of such a modem design with the real channel. The use of an advanced environment like GRAPE in combination with programmable hardware strongly encourages the fast prototyping and makes it hardly more expensive than simulation.

GRAPE's advanced use of run-time readable and writable parameters facilitates the monitoring, debugging, and fine-tuning of applications assigned to hardware platforms which can be exchanged easily and independently from the application. By using DSPs instead of workstation simulation the execution times of the algorithm are tremendously speeded up by a factor in the order of 10^2.

6. Acknowledgments

R. Lauwereins is a Senior Research Associate with the FWO. M. Adé is a postdoctoral researcher of the K.U. Leuven-DOC. J. Dalcolmo and A. Hein are postdoctoral researchers, P. Le Corre is a master of engineering at the K.U. Leuven. Research has been supported by Siemens Atea and the Flemish Government via the Flemish Institute for the Advancement of Scientific-Technological Research in Industry (IWT). This project has been partly made possible by an FWO Krediet aan Navorsers and is partly sponsored under Texas Instruments' Elite University Program. K.U. Leuven-ESAT is a member of the DSP Valley network.

References

[1] M. Adé, R. Lauwereins, and J. Peperstraete. Hardware-software co-design with GRAPE. In *Proc. 6th Int. Workshop on Rapid System Prototyping*, pages 40–47, June 1995. Chapel Hill, NC, USA.

[2] G. Bilsen, M. Engels, R. Lauwereins, and J. Peperstraete. Cyclo-static dataflow. *IEEE Transactions on Signal Processing*, 44(2):397–408, February 1996.

[3] R. Comerford and S. Tekla. Wired for interactivity. *IEEE Spectrum*, pages 21–28, April 1996.

[4] W. End. IEEE project 802.14 : Standards for digital convergence. *IEEE Communications Magazine*, pages 20–23, May 1995.

[5] L. Goldberg. Cable modems: The journey from hype to hardware. *Electronic Design*, pages 65–80, April 1996.

[6] R. Lauwereins, M. Adé, M. Engels, and J. Peperstraete. GRAPE-II: A system-level prototyping environment for DSP applications. *IEEE Computer*, 28(2):35–43, February 1995.

[7] R. Lauwereins, M. Adé, P. Vandaele, M. Moonen, and P. Schaumont. Prototyping quadrature amplitude modulation for two-way communication on CATV networks. In *Proc. of 7th Int. Conference on Signal Processing Applications and Technology ICSPAT*, pages 1570–1574, 1996. Boston.

[8] E. Lee and D. Messerschmitt. Static scheduling of synchronous data flow programs for digital signal processing. *IEEE Transactions on Computers*, C-36(1):24–35, January 1987.

CONSTRAINT ANALYSIS FOR DSP CODE GENERATION

Bart Mesman[1,2], Marino T.J. Strik[1], Adwin H. Timmer[1], Jef L. van Meerbergen[1] and Jochen A.G. Jess[2]

[1]Philips Research Laboratories, WAY4, Prof. Holstlaan 4, 5656 AA Eindhoven, The Netherlands
[2] Section ICS, Department of Electrical Engineering, Eindhoven University of Technology, The Netherlands

Abstract

Code generation methods for DSP applications are hampered by the combination of tight timing constraints imposed by the performance requirements of DSP algorithms, and resource constraints imposed by a hardware architecture. In this paper, we present a method to analyze resource- and timing constraints in a single model. The analysis identifies sequencing constraints between operations additional to the precedence constraints. Without the explicit modeling of these sequencing constraints, a scheduler is often not capable of finding a solution that satisfies the timing and resource constraints. The presented approach results in an efficient method of obtaining high quality instruction schedules.

1 Introduction

Digital Signal Processing (DSP) algorithms have the following characteristics: they allow parallel execution of a number of operations, periodic executions, and tight timing requirements. These algorithms can be represented using a signal flow graph (SFG), consisting of operations, data flow edges and sequence edges. The precedence constraints expressed in such a SFG combined with the tight timing requirements, already pose a problem for most scheduling methods targeting for efficient hardware. However, mapping the DSP algorithm on a hardware platform also introduces hardware constraints, which make the schedule problem even more difficult to handle. The hardware constraints from the mapping depend on the choice of hardware platform. We consider two types of hardware on which to map a DSP algorithm: ASIPs ([9], application domain specific hardware), and digital signal processors. ASIPs are tuned towards specific application domains and have become popular due to their advantageous trade-off between flexibility and cost. The fixed architecture of an ASIP introduces resource constraints for the instruction scheduler in the code generation process.

General purpose digital signal processors offer more flexibility than ASIPs.They also introduce resource constraints as a result of a fixed architecture. For reasons of hardware efficiency and code density, the instruction set of both an ASIP and a general purpose DSP is often irregular, excluding certain types of parallel execution. This introduces another kind of hardware constraint a scheduler has to deal with.

In order to obtain efficient code, it is necessary for a scheduler to take into account resource constraints, instruction conflicts, and timing constraints originating from the application. The combination of these types of constraints poses a problem for traditional scheduling techniques. Many scheduling heuristics have been proposed, each with an optimization criterion. They are however unable to cope with the interactions of different types of constraints. In this paper we therefore address the schedule problem as a feasibility problem and propose a solution strategy based on the analysis of all the constraints.

Section 2 discusses some related work, illustrating the difficulty of scheduling with tight constraints, and proposes a global solution strategy. In Section 3 we show how all the different constraints can be combined in the model of a precedence graph. In Section 4 we describe the method of analysis and the interaction with a simple off-the-shelf scheduler to obtain a schedule, and in Section 5 some results will be presented.

1080-1820/97 $10.00 © 1997 IEEE

2 Related work and contributions

As a result of the difficulty in solving the DSP scheduling problem, a range of heuristics, each with its own optimization criterion, have been proposed. Probably the most widely used scheduling heuristic in high-level synthesis [7] is the list scheduler [1]. This heuristic takes a fixed set of resources, and optimizes latency (completion time). A well-known heuristic that optimizes resources given a constraint on the latency and throughput is force-directed scheduling [2]. Furthermore, in order to obtain higher utilization rates for the resources, it is useful to let executions of different loop iterations overlap. When using a counter-based controller, like in the Phideo compiler [4], this can be done very efficiently. In a micro-coded controller, overlapping executions are realized by software pipelining [3], also called loop pipelining or loop folding. Such methods assume that the scheduler is able to handle resource conflicts between operations belonging to different loop iterations. The difficulty in handling these inter-iteration conflicts is illustrated with a small example in Figure 1. In this figure, a precedence graph of 5 operations is given (the arrows indicate a precedence relation). In order to meet the constraint of 3 clock cycles on the initiation interval (II), loop folding has to be applied (indicated by the arrow in Figure 1 b and c). Because folding introduces extra code, we do not want to fold more than once, which constrains the latency to 6 clock cycles. In Figure 1b the result of a list scheduler is shown. The left column contains the *time potential* (schedule time modulo II). The list scheduler greedily schedules A, B, and C as soon as possible (ASAP), and concludes that D cannot be scheduled. In Figure 1c a valid schedule is given. The key to obtaining this schedule is to postpone B one clock cycle relative to its ASAP value. However, most schedule heuristics lack the scope to postpone operations. The scope of a scheduler could be expressed as the schedule freedom (in clock cycles per operation) available to it. The scope of most scheduling heuristics is limited to the apparent schedule freedom (also called mobility), calculated as the difference between the ALAP (as late as possible) value and the ASAP (as soon as possible) value, based on the precedences between operations, but without taking the resource conflicts into account. In Figure 1, the apparent freedom is 1 clock cycle per operation. The reader can verify that the combination of precedence, resource, latency, and throughput constraints leaves no schedule freedom at all: the schedule in Figure 1c is the only possible schedule in 6 clock cycles.

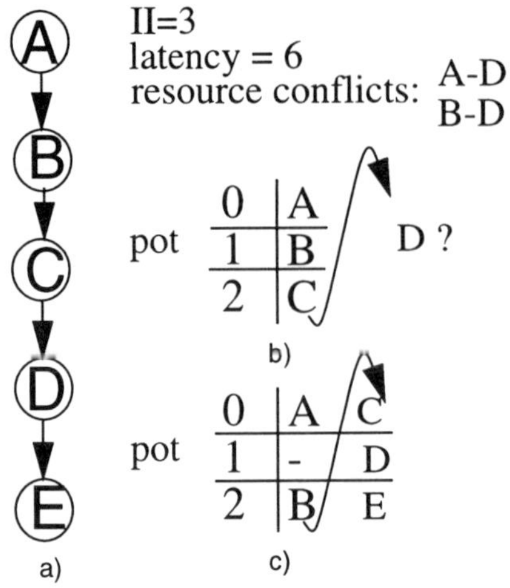

Figure 1 Example with loop folding. a) precedence graph b) list-schedule c) only feasible schedule in 6 clock cycles

The general problem statement for finding a feasible schedule, as depicted above, is as follows.

Problem 1: Given a (cyclic) signal flow graph (SFG), a set of resource conflicts, a latency, and an initiation interval (II), find a schedule.

In this context, resource conflicts include conflicts on execution units, bus access, memory access, and instruction conflicts. Note that two tasks have been ignored in the problem statement: finding the optimal initiation interval, and determining a resource-binding for the operations. We assume that instruction selection has been done (for example using method [12]) prior to the scheduling phase, thus providing a resource binding. The optimal II is found as follows. We start with a lower bound based on loop-carried dependencies [11] and available resources. When this II is not feasible, it is incremented by one clock cycle. Profiling suggests that the optimal II is usually only one or two clock cycles away from the lower bound.

A special case of problem 1 has been proven NP-complete in [8]. Because of this inherent difficulty and the limited success of scheduling heuristics on heavily constrained applications, we propose a solution strategy based on the interaction of analyzing the constraints, and making scheduling decisions. The problem statement for the analyzer is then as follows.

Problem 2: Given a SFG, a set of schedule decisions, a set of resource conflicts, a latency, and an initiation interval, reduce the schedule freedom as much as possible without excluding any feasible solutions.

In Section 4 we will use analyses to tackle Problem 2, and then show how these analyses can be used to tackle Problem 1.

Problem 2 is essentially constraint analysis. Not much work has been done in this field. In [5] the combination of resource and timing constraints is analyzed in a general way. Because it is difficult to extend this work to allow loop folding, it is not suitable for our specific schedule problem. Loop folding in the context of resource and timing constraints is considered in [6]. In this work, a bipartite matching formulation is used to analyze the matching of execution intervals of operations to execution intervals of resources. Reductions in the execution intervals are obtained by showing that some matchings can never be part of a complete matching. The bipartite-matching approach is based on the concept of execution interval. The approach taken in this paper is based on sequence relations between operations. In Section 5, we will compare these two approaches, and show that our approach is able to obtain further reductions in the schedule freedom after the analysis done in [6].

3 Modeling the constraints

The analyzer needs a uniform model of the DSP algorithm, which enables the expression of all constraints mentioned in the introduction. For this purpose, the model of a (directed) precedence graph is used [10]. It consists of a set V of vertices representing operations, and a set $A \subseteq V \times V$ of arcs, called precedence edges or sequence edges, representing precedence relations between operations. For $v \in V$ let $s(v)$ denote the start time of operation v. An arc (v_i, v_j) with weight w indicates that $s(v_j) \geq s(v_i) + w$. Two dummy operations are added to the precedence graph: a source and a sink. The source operation is always the first operation to execute, and the sink the last one. We will now show how constraints can be incorporated in this model.

- **Data dependencies**. The translation of the data dependencies in this model is straightforward: each dependency translates to an arc with weight equal to the execution delay of the producing operation.
- **Latency**. A constraint l on the latency is translated to an arc (sink, source) with $w=-l$, as illustrated in Figure 2. This is interpreted as $s(\text{source}) \geq s(\text{sink}) - l$, which is equivalent to $s(\text{sink}) \leq s(\text{source}) + l$, meaning the last operation may not be executed more than l clock cycles after the start of the first operation.

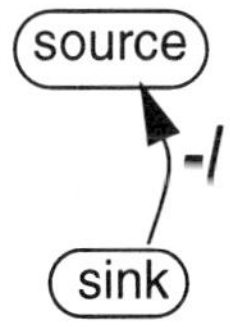

Figure 2 Modelling the

- **Microcoded controller and loop folding**. We assume that the architecture contains a microcoded controller. As a consequence, the same code is executed every loop iteration. This implies that a variable is written in the same register each iteration. When loop iterations overlap, we have to ensure that a variable is consumed before it is overwritten by the next production. Since subsequent productions are exactly II (initiation interval) clock cycles apart, a variable cannot be alive longer than II clock cycles. So the operation C that consumes a variable must execute within II clock cycles after the operation P that produces the variable. Just like the latency constraint, a necessary and sufficient translation to the precedence model is that for each data dependency (P,C) there is an arc (C,P) with $w = -II$.
- **Pipelined executions and multicycle operations**. Pipelined executions and multicycle operations can be modelled by introducing an operation for each stage of the execution. Subsequent stages are linked in time using two sequence edges as indicated in Figure 3. For multicycle operations, A and B occupy the same resource.

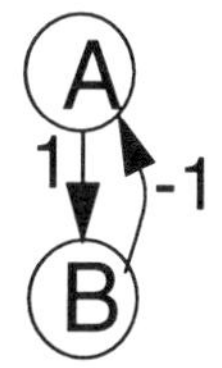

Figure 3 Modeling pipelined and multicycle operations

- **Scheduling decisions**. Scheduling decisions may take different forms. A timing relation between two operations can be directly translated to a sequence edge. When an operation v is fixed at a certain clock cycle c, we need two sequence edges as indicated in Figure 4

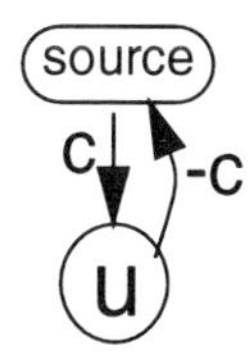

Figure 4 Modeling a schedule

- **Resource conflicts and instruction set conflicts**. We use method [6] to model instruction set conflicts as resource conflicts. To incorporate resource conflicts we need an additional formalism.

Let the function $\text{rsc}(v_i, v_j) : V \times V \rightarrow \{0,1\}$ be defined by

$$\text{rsc}(v_i, v_j) = \begin{cases} 1 \text{ if } v_i \text{ and } v_j \text{ have a conflict} \\ 0 \text{ otherwise} \end{cases}$$

then $\text{rsc}(v_i, v_j) = 1 \Rightarrow s(v_i) \neq s(v_j) \pmod{\text{II}}$.

4 Constraint analysis and schedule interaction

4.1 Constraint analysis

In this section we elaborate on the solution strategy for problem 1 and 2. First we focus on problem 2: The reduction of the schedule freedom based on the analysis of the constraints. This reduction is made explicit by adding sequencing constraints (edges). Such sequencing constraints are easily handled by all scheduling heuristics, so they effectively reduce the schedule freedom available to the scheduler. The sequence constraints enable the analyzer to observe the different constraints in a uniform model (precedence graph). By incorporating all the constraints in a uniform model we can efficiently explore the interactions between the different constraints. One can think of rules that express the interaction between different types of constraints in terms of a sequence constraint. We show two effective rules that involve the combination of resource conflicts, precedence, and timing constraints. The extension of our approach to incorporate register conflicts is illustrated by a simple rule that sequentializes variable lifetimes.

All the rules used in our approach use the concept of a path between operations.

Definition: A path of length d from operation v_i to operation v_j is a chain of precedences v_i -> v_k ->...v_l ->v_j that imply $s(v_j) \geq s(v_i) + \text{d}$

Definition: The distance $\text{d}(v_i, v_j)$ from operation v_i to operation v_j is the length of the longest path from v_i to v_j.

In the following examples a path is indicated using a dashed arc labelled with the length of the path.

The first rule presented below, affects the timing relation between conflicting operations. It is based on the fact that two conflicting operations cannot be scheduled at the same potential. The *time potential* associated to a time t is t mod II. So if the distance between these operations would cause them to be scheduled at the same potential, the distance has to be increased by at least one clock cycle.

Rule 1: If $\text{d}(v_i, v_j) \equiv 0 \pmod{\text{II}}$ and $rsc(v_i, v_j) = 1$, we can add a sequence edge (v_i, v_j) with weight $\text{d}(v_i, v_j) + 1$.

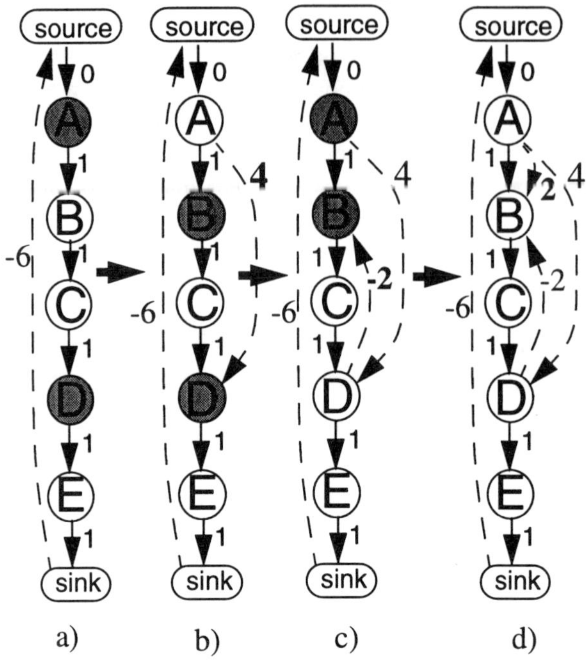

Figure 5 Derivation of a schedule for Figure 1

This rule will help us to solve the schedule problem in Figure 1. From the discussion in Section 2 we concluded that the difficulty of the schedule problem is to put a gap of one clock cycle between A and B. So our goal is to derive that d(A,B) = 2. In Figure 5 this derivation is given. Figure 5a represents the precedence model of Figure 1 a. In Figure 5a we see a path A->B->C->D of length 3 ≡ 0 mod II from A to D. According to rule 1 we can add a sequence edge A->D of weight 4 because A and D have a resource conflict. This edge is drawn in Figure 5b. Next there is a path D->E->sink->source->A->B of length 1+1-6+0+1 =-3 clock cycles. Because of the resource conflict D-B, this length has to be increased by one clock cycle. This gives a sequence edge D->B of weight -2, as given in Figure 5c. We conclude by finding a path A->D->B of length 4-2=2 clock cycles. In Figure 5d the associated sequence edge (A,B) of weight 2 is explicitly drawn. The precedence relations now completely fix the schedule.

The second rule we present in this paper is more complicated, and involves symmetry in the precedence graph. Consider the small piece of precedence graph depicted in Figure 6. The distance from A to D is two clock cycles. However, B and C have to be ordered because they have a resource conflict, and both possible orderings will result in d(A,D)=3. Rule 1 will not help us here. In DSP-algorithms, this type of symmetry is a very frequent phenomenon and has to be dealt with in the analyzer. Rule 2 will cope with this issue in the context of loop folding.

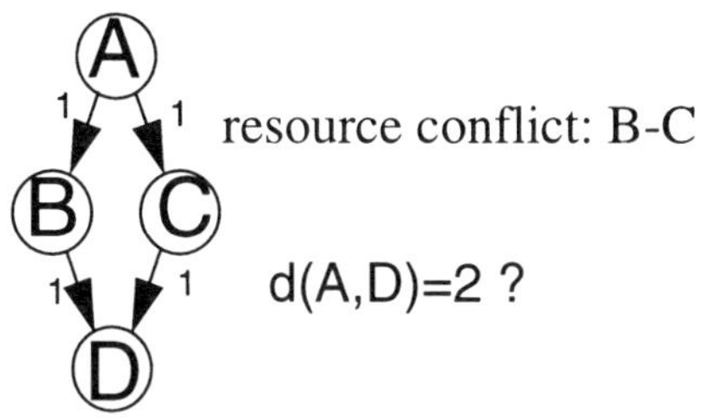

Figure 6 Too much apparent mobility due to symmetry

Rule 2: For each pair of operations v_i and v_j such that $rsc(v_i, v_j) = 1$, if there is an operation p such that $d(p, v_i) \equiv d(p, v_j) \pmod{II}$, and an operation s such that $d(p, v_i) + d(v_i, s) = d(p, v_j) + d(v_j, s)$, we can add a sequence edge (p, s) with weight $d(p, v_i)+d(v_i, s)+1$

In Figure 6, operation p is A, and s is D. We may add a sequence edge (A,D) of weight d(A,B) + d(B,D) + 1 = 3.

In Figure 7, the symmetry is of a slightly different kind. As can be seen in the ASAP schedule, the only way the minimum distance of 4 clock cycles from A to H can be realized, is to schedule operations B and G at the same potential. Because B and G have a resource conflict, the distance from A to H is not 4, but 5 clock cycles. This also follows from rule 2: d(A,B) = 1 and d(A,G) = 3, so d(A,B) $\equiv$ d(A,G) (mod 2). Also d(A,B) + d(B,H) = 4 and d(A,G) + d(G,H) = 4, so we may add a sequence edge (A,H) of weight 5.

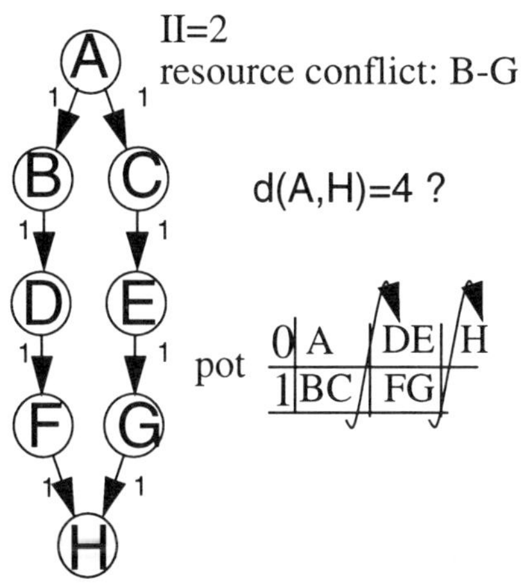

Figure 7 Applying rule 2 with loop folding

The next rule (rule 3) considers register conflicts.

Rule 3: Let variable v1, produced by operation p1 and consumed by c1, and variable v2, produced by operation p1 and consumed by c1, reside in the same register. If $d(p_1, p_2) \geq 0$ we can add a sequence edge (p_1, p_2) with weight 0.

Rule 3 is illustrated in Figure 8. The variables v1 and v2 are bound to the same register. This is often a priori known because most general purpose DSPs have special purpose registers and flag registers. As a result, the binding of some variables is fixed when special purpose registers or flags are used. If there is a path of positive length from P1 to P2, then the whole lifetime of variable v1 has to precede the lifetime of v2. This is made explicit by adding a sequence edge from the consumer C1 to the producer P2.

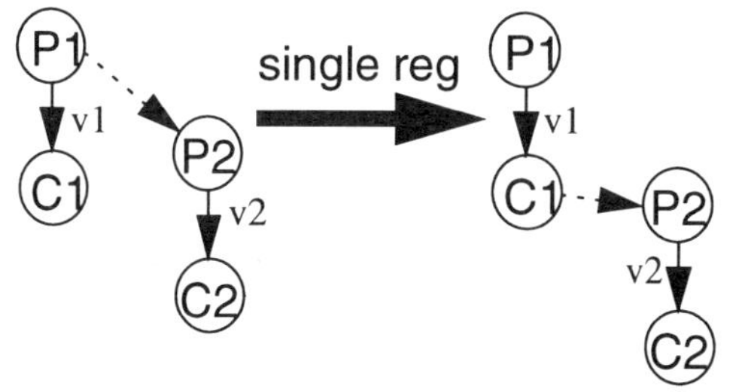

Figure 8 Rule 3 for sequentializing variable lifetimes

4.2 Complexity

In our implementation the longest path between each pair of operations is administrated. So if a new edge is added, the impact on the current longest paths has to be calculated. Therefore the complexity of adding a sequence edge is the dominant factor in run time. This complexity is essentially determined by the number of paths that need to be updated as a result of the new sequence edge. Because we are only interested in the longest paths found so far, the number of updates equals V^2 worst case. In most cases, the addition of a sequence edge will affect a few paths. In cases where a lot of paths need to be updated, the reduction in schedule freedom will also be substantial.

An upperbound on the number of path updates (as a result of adding a sequence edge) can be derived as follows. A path can have a length between $-l$ and $+l$, where l is the constraint on the latency. Because a path is updated only if its length is increased (by at least one clock cycle), the number of times a path can be updated is at most $2l$. Since the maximum number of paths we keep track of, equals V^2, the number of path updates can be at most $2l \cdot V^2$. A single path update takes constant time, so the run-time of the analysis is polynomially bounded.

4.3 Interaction with an instruction scheduler

Rules 1 to 3 enable us to tackle Problem 2, mentioned in Section 2. Problem 1 is handled by interacting between the analyzer and a scheduler. For this interaction to work we need to address two issues:

- The analyser can handle schedule decisions. Schedule decisions can be modelled as indicated in Section 3.
- The scheduler can take the analysis results into account. These results are in terms of sequence edges and thus express precedence. The least one can expect from an instruction scheduler is to take precedences into account.

Furthermore, we should not forget that we are dealing with an NP-hard problem. So it is very well possible that the analyzer is unable to prevent a scheduler from making an infeasible decision, thus forcing us to backtrack.

5 Results

Three experiments are reported in this section. In two cases, our approach (Constrained Analysis and Scheduling, or CAS) is compared to the approach of [6] using a bipartite schedule graph (BSG). The first experiment shows the effect of loop folding. The second experiment illustrates the effect of register binding. The last experiment, an example taken from an actual design, is used to show the run time in a practical application. With respect to the resource constraints, only rules 1 and 2 have been used. In the last two experiments, also some rules are used for considering register conflicts. The quality of the analysis is measured by the reduction in schedule freedom.

The first example is a radix-2 butterfly, shown in Figure 9. Because the multiplication is a multicycle operation, it is split in two stages m1 and m2, making a total of 10 operations. In Table 1 the result of the analysis is shown, expressed in the average schedule freedom per operation in clock cycles. For the non folded schedule, the reduction in schedule freedom is exactly the same for BSG and CAS. However, folding the schedule increases the reduction for the BSG analysis only slightly, whereas the CAS analysis shows that only one operation (r0) has any schedule freedom left. For this example there is no gain in accuracy by combining the results of BSG analysis and CAS analysis. The run times for both BSG and CAS are negligible.

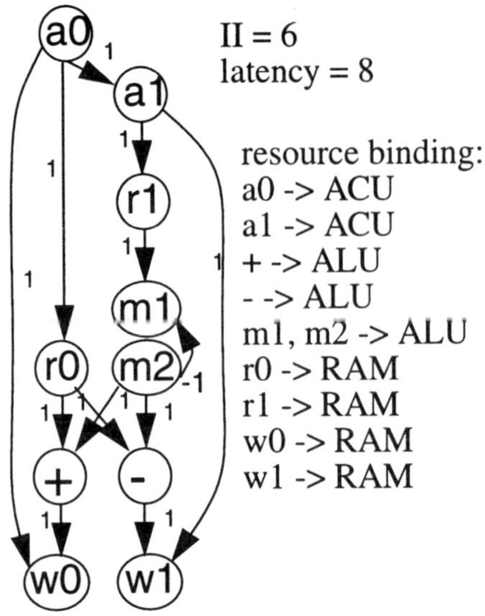

Figure 9 Radix-2 butterfly used in first

Table 1 Average schedule freedom for radix-2 butterfly

	ASAP-ALAP	BSG	CAS
non folded	1.2	.7	.7
folded	1.2	.5	.1

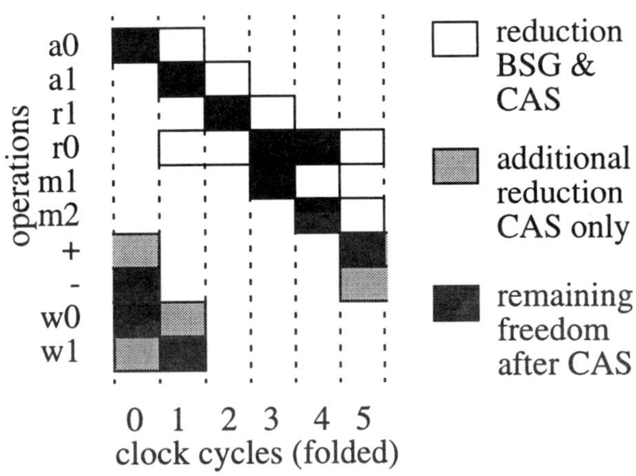

Figure 10 Schedule freedom per operation

The execution intervals for each operation are given in Figure 10 for the folded schedule. In this figure, the time is represented in clock cycles at the horizontal axis. The operations are enumerated vertically. The white area represents the reductions obtained by both BSG and CAS analysis. For example, the execution interval of operation r0, based on ASAP and ALAP is [1, 5]. BSG and CAS are able to reduce this interval to [3,4]. The grey area is the reduction obtained by CAS, that BSG was not able to find.

Notice in this figure how reduction techniques such as CAS and BSG prevent a greedy scheduler from making a wrong schedule decision. A greedy scheduler would schedule operation r0 in clock cycle 1, leaving no room for w1 in clock cycle 7 (potential 1), which is the only feasible start time for w1. Both CAS and BSG detect that r0 can not be scheduled in clock cycle 1. However, BSG is unable to detect that operation + must precede operation -. A greedy scheduler can easily go wrong by letting operation - precede operation +.

The second experiment concerns an IIR filter of 23 operations, including fetching the coefficients and data from memory. The minimum latency is 10 clock cycles, which equals the latency constraint.

Table 2 Average freedom for IIR filter

	ALAP-ASAP	BSG	CAS	com-bined	CAS, 1 decision	CAS, reg
non folded	2.70	1.61	1.83	1.52	.68	.30
folded	2.70	1.61	1.74	1.43	.56	.13

In Table 2 the result of the analysis is shown. BSG analysis is capable of deducing some reductions CAS can not find, and vice versa. As can be seen in Table 2, combining the analyses provides larger reductions than both analyses separately. In the sixth column, the remaining schedule freedom is shown when CAS analyzes the first schedule decision a greedy scheduler could make (operation 23 at clock cycle 0). It is clear from the numbers in the fifth column that substantial reductions can be made not only before scheduling, but also during scheduling. This observation strengthens the idea of an interaction between the analyzer and the scheduler. The last column shows the reduction in schedule freedom when rules are applied in CAS that consider register conflicts. These register conflicts are the result of a register binding. Using our approach, we were able to prove that we used the least possible amount of registers in order to satisfy the timing constraints. It is clear from Table 2 that the best reduction is obtained by considering register conflicts as well.

Table 3 Average schedule freedom for FFT loops, CAS only

	ALAP-ASAP	resource con-flicts	register conflicts as well
FFT (40 op's)	4.40	3.41	.71
Radix (80op's)	3.81	2.08	1.38

The third experiment includes only CAS, and concerns two loops of FFT algorithms. The first loop (FFT) contains 40 operations, has a minimum latency of 13 clock cycles, and needs to be folded at least three times to realize an initiation interval of only 4 clock cycles. The second loop (Radix) contains 80 operations, has a minimum latency of 11 clock cycles, and is folded twice to realize an initiation interval of 4 clock cycles. In Table 3 the results are shown. The reduction obtained on FFT by taking into account the register conflicts, is considerably more than in the second experiment. The run time of this experiment using the most extensive analysis, is less than a second on a HP 9000/735.

6 Conclusions

In this paper, we presented an approach for code generation based on the analysis of precedence, timing and resource constraints. We also hinted at the way register conflicts are considered in CAS. By making all constraints explicit in a graph model and calculating the longest paths, we are able to see the interaction between the different constraints, and compute the effect on the schedule freedom available to a scheduler. The results in Section 5 show that the reduction in schedule freedom obtained by CAS is substantial. With loop folding we are able to obtain more accurate reductions than the BSG approach in [6]. Furthermore, in CAS it is easy to incorporate rules, such as shown in Section 4, that take register conflicts into account. BSG is not able to handle this type of conflict. We also showed that the obtained reductions really prevent a greedy scheduler from making a wrong decision. We have illustrated how the interaction with a scheduler works by showing the reduction in schedule freedom CAS can make as a result of a schedule decision. We conclude that analysis tools such as CAS are needed in order to obtain a feasible schedule when facing resource constraints, register constraints, and tight timing constraints.

Acknowledgments

We would like to thank Paul Lippens and Koen van Eijk for their support and constructive discussions.

References

[1] D.E Thomas, E.D. Lagnese, R.A. Walker, J.A. Nestor, J.V. Rajan and R.L. Blackburn, "Algorithmic and register-transfer sysnthesis: The systems architect's workbench", Kluwer Academic Publisher, 1990

[2] P.G. Paulin and J.P. Knight, "Force-directed scheduling for the behavioural sythesis of ASIC's", *IEEE Transactions on Computer-Aided Design*, pp. 680-685, June 1989

[3] G. Goossens, J. Vandewalle and H. De Man, "Loop optimization in register-transfer scheduling for DSP-systems", *Proc. 26th DAC*, pp. 826-831, 1989

[4] W.F.J Verhaegh, P.E.R. Lippens, E.H.L. Aarts and J.L. van Meerbergen, "Multidimensional periodic scheduling: A solution approach", *Proc. ED&TC 1997*, pp. 468-474, March 1997

[5] W.P.M. Nuijten "Time and resource constrained scheduling, a constraint satisfaction approach", Ph.D. thesis, Eindhoven University of Technology, 1994

[6] A.H. Timmer, M.T.J. Strik, J.L. van Meerbergen and J.A.G. Jess, "Conflict modelling and instruction scheduling in code generation for in-house DSP cores", *Proc. 32nd DAC*

[7] M. C.SJ. McFarland, A.C. Parker and P. Camposano, "Tutorial on High-level synthesis", *Proc. 25th DAC*, pp. 330-336, 1988

[8] M.R. Garey and D.S. Johnson, "Computers and intractability: A guide to the theory of NP-completeness", Freeman, 1979

[9] R. Leupers, W. Schenk and P. Marwedel, "Microcode generation for flexible parallel architectures", *Proc. Working Conf. Parallel Archit. and Compil. Techn.*, North-Holland, 1994

[10] D.C. Ku and G. De Micheli, "High-level synthesis of ASICs under timing and synchronization constraints", Kluwer Academic Publishers, 1992.

[11] R. Reiter, "Scheduling parallel computation", *Journal of the ACM*, vol.15, pp. 590-599, 1968

[12] C.Liem, T.May and P. Paulin, "Instruction-set matching and selection for DSP and ASIP code generation", Proc. ED&TC pp. 31-37, Paris, Feb. 1994

An Efficient Model for DSP Code Generation: Performance, Code Size, Estimated Energy

Catherine H. Gebotys
Department of Electrical and Computer Engineering
University of Waterloo, Waterloo, Ontario,Canada N2L 3G1
cgebotys@optimal.vlsi.uwaterloo.edu

Abstract

This paper presents a model for simultaneous instruction selection, compaction, and register allocation. An arc mapping model along with logical propositions is used to create an optimization model. Code is generated in fast cpu times and is optimized for minimum code size, maximum performance or estimated energy dissipation. Code generated for realistic DSP applications provide performance and code size improvements from 1.09 up to 2.18 times for the TMS320C2x processor compared to previous research and a commercial compiler. In all examples up to 106 instructions are generated in under one cpu minute. This research is important for industry since DSP code can be efficiently generated with constraints on code size, performance, energy dissipation.

1.Introduction

High reliability, high performance and low cost are forcing higher levels of integration for VLSI embedded systems. As DSP applications are rapidly growing more complex, some designers are moving from full custom digital circuitry to programmable processors or in-house cores to obtain lower risk solutions. The DSP core is a DSP processor that can be reused and combined with program/data memory, dedicated logic, plus ASICs, and incorporated onto a large silicon chip, providing a cost efficient and flexible solution for many typical embedded applications requiring low power and high reliability. These systems demand small code size and high performance. Due to increasing complexities, high level compilation is a necessity. However the biggest drawback to both DSP processors or DSP core use is the code generation.

The use of conventional code generation techniques and even compilers specifically designed for commercial DSP processors produce very inefficient code [8,4]. There are many more limitations placed upon code generation for the DSP processor than for the general purpose processor. The difficulty arises from non-homogeneous register sets, small number of very specialized registers, very specialized functional units, restricted connectivity, limited addressing, and highly irregular datapaths[8]. For example specialized functional units such as multiplier-accumulators and address calculation units are typically found in these architectures. Instructions take operands and store results of computations in well defined registers with limited connectivity. Instructions are highly interdependent and involve the use of modes, for example the mode classes typically involve product-shift and sign-extension. Some instructions will function differently depending upon the current mode setting and extra instructions to reset the mode are typically required. Limited addressing modes and the use of address registers are also typical. Code generation for DSP processors must meet tight timing constraints dictated typically by DSP throughput, and meet tight resource constraints. Given that these DSP processors must meet these requirements using very small code space (all on chip ROM), the code generation problem is a very difficult one[10]. Typically DSP processors are difficult to use and require long product development times even though the program being developed may be stored on chip in less than 1K program ROM. The need for decreasing time to market, development costs, and maintenance costs, demands the use of high level language compilation. All of these factors imply several challenges in writing efficient code generators for such DSP processors. This is even more difficult for the design of in-house DSP processor cores since retargetable compilation is also required.

2. Problem Description and Related Work

The following problem, problem 1 given below, is an important part of the code optimization problem that will be studied in this paper. For simplicity let us assume that an algorithm to implement the application has already been assigned based upon accuracy required, low energy implementation, etc.. The algorithm is composed of basic blocks, which are given as a partially ordered list of code operations. For the problem definition below we assume that there is one target DSP processor or core defined with an instruction set architecture (instruction set architecture).

Problem 1.

Assume we are given a sequence of operations, represented by an ordered list of operations. The objective is to select instructions, compact code, and allocate registers for the target processor. The set of operations is mapped into a set of instructions, which is optimized for code size, performance, or energy dissipation.

Researchers have shown that code requiring a minimum number of cycles can be generated for expression trees[1] and an extension for general DAGs was researched in [2], however both techniques do not consider instruction-level parallelism. Researchers in [3] use graph based technique to perform code compaction, whereas in [4] a branch and bound scheduler and a heuristic search scheduler for the TMS320C2x processor are used to try to reduce the accumulator spilling and mode cost. A large integer linear programming (ILP) model was researched in [5] that performed simultaneous instruction selection, code compaction and register allocation. Unfortunately it was too complex for realistic processor representation.

Recent research in integer optimization has found that large subclasses of logical inference problems can be solved as a relaxed linear programming problem (LP) [6]. In particular if the system is composed of horn clauses[7] the optimal integer solution can be obtained from the relaxed linear programming problem solution (where binary variables are replaced by continuous positive variables with upper bounds equal to one). In the solution to the linear programming problem, any variable which does not have an integer value can be rounded up or down to the nearest integer value. It has been proven that this final integer solution is always optimal[7]. Researchers have also found that by rewriting constraints as horn clauses the computational time for solving the problem can be greatly reduced. Experience has shown that many logical inference problems can be solved as relaxed LP problems even if they are not horn clauses[7].

In this manuscript a new approach is presented to solve problem 1, DSP code generation. Unlike previous research, the target DSP processor is defined using arc mappings and propositional logic which is then automatically translated into mathematical programming constraints to form the optimization model. The basic block components of the application are transformed into a sequence of operations, which is input to the model. The model generates code for the target processor. This process supports retargetable compilation for processors whose instruction set architecture can be described using propositional logic. The next section will outline the assumptions and terminology to be used in the rest of the paper.

3. Assumptions and Notation

The following terminology will be used in this paper:

o is an operation in the basic block of an application.

$o1 \rightarrow o2$ is an arc where data produced by operation $o1$ is used as input for operation $o2$.

s is an integer valued parameter representing the set of instructions that an operation o may be mapped into.

$T(o)$ is the time that the operation o is scheduled at.

$P_{o,s}$ is a proposition representing the mapping of operation o into the set of instructions s.

$x_{o,s}$ is a binary variable representing the proposition $P_{o,s}$, when the proposition is true, the binary variable equals 1 otherwise it equals 0.

P_o is a proposition when true means that there is some parallel instruction that can be obtained by merging an instruction in the set of instructions for operation o with another instruction in the set of instructions for the next operation, $o1$, where $T(o)=T(o1)-1$.

p_o is the binary variable representing the parallelism proposition P_o.

$End(o,s)$ is the end resource being used to store the data output from this operation, o. For example the output data may be stored in a register or in memory (and accessed as a memory operand by another operation).

$E(o)$ is a parameter that identifies the type of arc for operation o. For example if $o \rightarrow o2$, where o is a multiplication and $o2$ is an addition then the type of arc for o, $E(o)$, is a *multiply-add* arc.

$cycles(o,s)$ is the number of cycles required to execute the set of instructions s for operation o.

$savings(o)$ is the number of cycles saved due to implementing a parallel instruction for o and $o2$, where $T(o)=T(o2)-1$.

$memaccesses(o,s)$ is the number of memory accesses performed for the set of instructions s for operation o (including memory operand accesses and the instruction accesses).

$msavings(o)$ is the number of memory accesses not required anymore due to implementing a parallel instruction for o and $o2$, where $T(o)=T(o2)-1$.

$codesize(o,s)$ is the number of instructions in the set of instructions s for operation o.

$csavings(o)$ is the number of instructions not required anymore due to implementing a parallel instruction for o and $o2$, where $T(o)=T(o2)-1$.

The rest of the paper will illustrate the model, constraints, and generated code results using the TMS320C2x processor[15]. The partial instruction set architecture for this processor is shown in table 1 (where || *Instr* refers to parallel instructions). This popular DSP processor has been described as having a very restrictive type of instruction level parallelism, making compaction a nontrivial task[8]. The optimization-based techniques presented in this paper provide a fast approach to generating optimized code. It is a retargetable model that can be used for modeling other DSP processors as well. The next section will provide an introduction to propositional logic and how it can be used to describe instruction set architecture constraints and mapped automatically into constraints of an ILP. The following sections, illustrate how arc mappings and ILP constraints are defined. Examples will be used to illustrate how optimization for code size is performed. Finally examples will be presented to show that new energy optimized solutions are produced unlike approaches using previous research or existing commercial compilers.

Table 1. Partial ISA TMS320C2x

Instr	Definition	II Instr	Definition
ADD m	a ← a + m	LTA m	LT m, APAC
APAC	a ← a + p	LTP m	LT m, PAC
SUB m	a ← a - m	LTS m	LT m, SPAC
SPAC	a ← a - p	MPYA m	APAC, MPY m
MPY m	p ← t * m	MPYS m	SPAC, MPY m
LAC m	a ← m		
LT m	t ← m		
PAC	a ← p		
SPL m	m ← p		
SACL m	m ← a		

† m: memory operand, special registers: a,p,t

4. Introduction to Propositional Logic

In general logical expressions can be used to model qualitative information about a processor or system. Propositions and propositional logic expressions can be automatically translated into binary variables and linear constraints. Specifically logical propositions are replaced by binary variables, where true and false are equivalent to binary values of one and zero. Reasoning can be shown to be equivalent to solving a mixed integer linear programming problem. Although there are many advantages to using a ILP model instead of a production system, the biggest drawback is computational time, since solving a ILP problem is NP-complete[9]. However if the system is completely composed of horn clauses it can be solved by a LP solver and a rounding technique. Several examples will be given below to illustrate horn clauses and their translation into constraints of a ILP problem.

Horn clauses are logical expressions that have at most one conclusion. Specifically horn clauses are disjunctions with not more than one non-negated term. For example if A is true and B is true then C is true can be expressed as:

$$\text{IF } P_a \text{ AND } P_b \text{ THEN } P_c$$

which can be translated into

$$\text{NOT } (P_a \text{ AND } P_b) \text{ OR } P_c$$

$$(\text{NOT } P_a) \text{ OR } (\text{NOT } P_b) \text{ OR } P_c$$

In this last disjunction, there is only one non-negated term (the last term, P_c), so it is a horn clause. One can automatically translate any logical proposition into an ILP constraint using binary variables. For example this last disjunction can be translated by using the following binary variables: x_A, x_B, and x_C to represent the propositions P_A, P_B, P_C respectively and write the following constraint:

$$(1-x_A)+(1-x_B)+x_C \geq 1$$

5. Methodology and Modeling

This section will briefly describe the methodology for code generation. Given an application described in a high level language (such as 'C') and a DSP processor to run the application, the following methodology maps the application into a set of instructions to be executed by the DSP processor. First the application is scheduled based upon the parallelism of the targeted DSP architecture (for example scheduling operations to make use of multiple functional units would be very important for many parallel DSP architectures). Next detailed instruction mapping is performed by first generating a code generation model. Arc mappings are developed and the logical propositions detailing constraints of the targeted instruction set architecture are developed. A model is then automatically generated from these logical propositions representing the ILP model for code generation. The high level language describing the application is decomposed into a series of basic blocks which are then optimized through the code generation model for performance, code size and finally estimated power. After the code is generated (which assumes direct memory addressing), the data layout is performed using existing[13] or extended techniques[11] followed by an optimization approach to address generation (using auxiliary registers) such as [12]. In this paper for illustration purposes we will use the TMS320C2x processor as the target processor. The partial instruction set architecture is shown in table 1. The TMS320C2x architecture is a non-homogeneous register architecture that contains special registers *a,p,t*, and has multiplier and alu functional units. The next two sections will illustrate how the model is generated from a target processor.

5.1. Arc Mappings

To model the target DSP processor in our model we first describe the arc mapping process. Any basic block from an application can typically be represented by a data flow graph, where each operation is a node and arcs define the transfer of data output from one operation to the input of another operation. In this model each arc can be mapped into different sets of instructions, each representing the path taken by the data value in the architecture. The path for example may be composed of register to memory or memory to register types of data transfer.

The arc mapping process is to determine for each type of arc the total number of different paths that could occur and identify each one as a set, s. For example in figure 1a) there is a multiplication operation *, which sends data to an addition operation +. We identify this type of arc as a *multiply-add* arc. This means that the data produced by the multiplication operation will be input to an addition operation. Figure 1a) shows the four possible paths for transferring the result of the multiplication operation to an input of the addition operation in the TMS320C2x processor. For s=1 the output of the multiplication is stored in register p until it is used by the addition operation. No extra instructions are required for this path. In s=2 the data is transferred from the p register into the a register right after the multiplication, using instruction *PAC*. In the remaining types it is transferred from the p register to memory (for example using instruction *SPLm*) and when it is required for input to the addition operation it is a memory operand (s=3) or loaded from memory into the a register (s=4) (for example using instruction *LACm*). Figure 1b) shows the three mappings for an *addition-subtraction* arc.

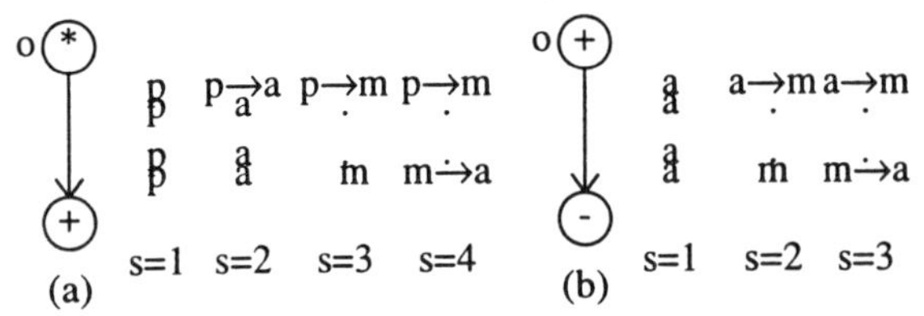

Figure 1. Example of arc mappings for *→+ in a) and +→– in b) for the TMS320C2x.

5.2 Instruction Set Architecture Constraints

Once the arc mappings are defined for the target processor, the constraints of the instruction set architecture of the DSP processor are defined using logical propositions and translated into a code generation model. The model takes the scheduled data flow graph (representing basic blocks of the application) as input and generates code as shown in figure 2. The example in figure 2 is the complex multiply ($cr = ar*br + ai*bi$, $ci = bi*ar + ai*br$). The corresponding scheduled data flow graph corresponding to a sequence of operations is shown at left hand side of figure 2. The final output from this model is shown at the right hand side. The logical propositions or the ILP constraints will also be given. Apart from basic model constraints the remaining processor specific constraints which look after conflicts in resources, register allocation, memory transfers, etc are horn clauses. Examples of some of the constraints for the code generation model will be provided below.

The basic formulation of the objective function to be minimized is a sum of costs. In particular the code size model is a direct sum of the number of instructions generated. The performance model is the actual number of cycles required to run the code and again a function of the instructions generated. The estimated energy model is a function of how many memory operands are used (where we account for the number of memory accesses from instruction memory and data memory).

Execution Time =

$$\sum_{o,s} cycles(o,s)x_{o,s} - \sum_{o} savings(o)p_o$$

Code Size =

$$\sum_{o,s} codesize(o,s)x_{o,s} - \sum_{o} csavings(o)p_o$$

Estimated Energy =

$$\sum_{o,s} memaccesses(o,s)x_{o,s} - \sum_{o} msavings(o)p_o$$

The simplest constraint is shown below which ensures that each operation must be assigned to only one set of instructions s.

$$\sum_{s} x_{o,s} = 1$$

The remaining constraints will be expressed using propositional logic to illustrate their representation as horn clauses.

The constraint that each input to an operation must be stored in a different resource, can be represented as : *for* $o1 \to o$ *and* $o2 \to o$*, where* $End(o1,s1)=End(o2,s1)$ *then*

IF $P_{o1,s1}$ THEN NOT $P_{o2,s2}$.

and translated into a mathematical ILP constraint using the transformations described in section 4. Similarly for the TMS320C2x, exactly one input to an adder must be in resource register a is formulated as follows: *for* $o1 \to o$ *and* $o2 \to o$*, where* o *is an addition operation,* $End(s1)=p$, $End(s2)=m$

IF $P_{o1,s1}$ THEN NOT $P_{o2,s2}$.

The general resource conflict constraints are illustrated in figure 3. In figure 3a) the operation * cannot be mapped into type s=1 (which uses register p) since operation *1 will always store it's output in register p. Similarly in figure 3b) operation + cannot be mapped into type t=1 due to conflict with resource a. This type of conflict constraint can be represented by, $o1 \to o2$, $T(o1) \le T(o3) \le T(o2)$, *where operation* $o2$ *uses a resource (such as the* a or p register) that is also being used to store the intermediate value generated by $o1$ *for* $o3$

IF $P_{o1,s1}$ AND $P_{o2,s2}$ THEN NOT $P_{o3,s3}$

Finally code compaction constraints are used to map to parallel instructions. These constraints are also horn clauses that restrict the parallel proposition P_o. For example if we have an addition operation that does not have any memory operands followed by a multiplication operation then we can use the *MPYA m* instruction to save on performance and code size (see table 1). Again this relates to two sequential operations , *where* $o1$ *is addition,* $o2$ *is multiplication,* $T(o1)=T(o2)-1$ $o11 \to o1$*,*$o12 \to o1$ *and* $End(o11,s1)=m$ *or* $End(o12,s2)=m$

IF ($P_{o11,s1}$ OR $P_{o12,s2}$) AND $P_{o1,s}$ AND $P_{o2,s}$ THEN NOT P_{o1}

This last constraint can easily be reformulated to support *mac* instructions in other processors, for example the M56000, where the multiplication can be chained with the addition operation.

This model also supports operations which transfer data to more than one operation. In this case a variable per operation is used to map to sets of instructions. For example input data to a multiplier can be a memory operand or loaded into the *t* register. If the same input data is used by more than one operation a second load into the *t* register (*LT m*) is not required. An example of this optimization is shown in figure 2 for input data *bi* which is stored only once in the *t* register, so the multiplication *bi*ar* does not require a *LT m* instruction. The model has also been extended to simultaneous scheduling. In this case the variable $x_{o,s}$ is replaced by $x_{o,t,s}$ where *t* refers to the time (or control step) that operation *o* is scheduled at. Precedence scheduling constraints are added and existing constraints are extended for scheduling.

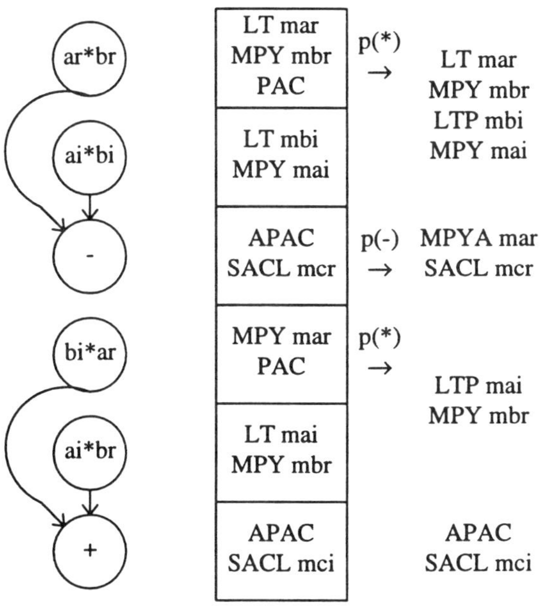

Figure 2. Example of mapping from a high level data flow graph into sets of instructions, and compacted code. These mappings are done simultaneously.

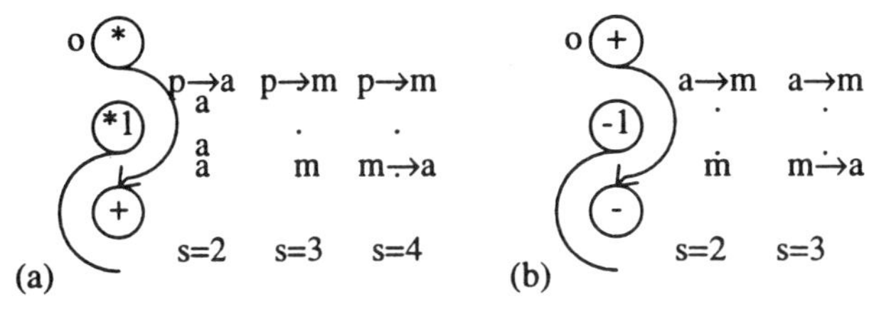

Figure 3. Example of conflict with resource *p* and *a* in a) and b) respectively, requiring the use of *s*=1 for both cases to be illegal.

6. Experimental Results

Several DSP examples are used to illustrate this methodology. Although any operation sequencer or scheduler can be used in comparison with previous research the same sequence of operations was used when we compared our results to previous research in order to make the comparison fair. All the code generated using the technique described in this paper has been verified through simulation with TI compiler generated code. The *cmul* example was taken from [11] and the *dfg* example was taken from [2]. The *lms* is a least means square algorithm and the *hp*1 and *hp*2 are high pass filter in direct form I and direct form II. The *fft* is a fast fourier transform taken from [16]. Although performance, code size or estimated energy can be optimized, the results in tables will provide code size or execution time results (or total number of cycles), since for TMS320C2x with on chip memory these are equivalent for basic blocks. Performance improvement is calculated as a [old # of cycles]/[new # of cycles] times improvement. All solutions are reported for the GAMS/Cplex LP/ILP solver[14] on a 133MHz Pentium PC.

Table 2 compares with the results of the TI C compiler generated code (set at the highest level of optimization) with the optimization model presented in this paper. To make a fair comparison all auxiliary register related instructions ie. computations/loads/etc were not included in the count. The code generated therefore assumes direct memory addressing and is only comparing actual computations related to the operations in the flow graph. For example the *cmul* example is that shown in figure 2 which uses 10 instructions (without counting indirect addressing-related instructions). The sequence of operations used for the *cmul* example is same as that used in [11] in order to compare with their code compaction technique which required 12 instructions.

Table 3 shows the size of the ILP models and the cpu times required to generate the solutions. When simultaneous scheduling was not performed, and a sequence of operations was input to the model, all solutions could be obtained from one LP solution. To illustrate the model performance the extension to simultaneous scheduling, instruction selection, register allocation and compaction was performed for the *cmul* example in only 2.3 cpu seconds, see second last row of table 3 (after searching through several nodes of the branch and bound tree). The code compactor in [11] (which does not perform instruction selection and register allocation) uses an ILP solution and runs on a RS/6000. A final run was also performed to see if the optimization model could improve upon the result of the sequencing of operations to optimize performance. In this case a cutoff value is used and the model for full scheduling, instruction selection, compaction and register allocation is solved. For the *dfg* example sequenced operations were obtained from the analysis in [2]. Using 21 instructions as a cutoff the ILP with simultaneous scheduling was solved and after 1042 cpu seconds it determined that the problem was integer infeasible. This proved that the code generated from the ILP model in this paper was in fact optimal.

Since the TI 'C' compiler is an industrial compiler a complete comparison of final code generated was performed with our code generation model presented in this paper in combination with the address optimization technique in [12]. These results are shown in table 4 (where *O* refers to the number of instructions used to generate computations of basic block alone and *A* refers to instructions used for address computations alone). Minimization of the number of memory accesses, to estimate energy optimization, was also performed. In the *fft* example an optimal solution was obtained in 1 cpu second (which was in fact the same as the optimal performance solution, shown in table 2).

Table 2. Performance Comparison

Example	# of Cycles/Code Size Compiler/Prev Res	Optim	Times Impr
cmul	15/12	10	1.50/1.20
dfg	24	21	1.14
hp1	48	22	2.18
hp2	48	30	1.60
lms	43	26	1.65
fft	108	99	1.09

Table 3. CPU times and Model sizes

Example	ILP Size Var	Eqn	CPU sec
lms	112	149	0.3
hp2	124	171	0.1
hp1	123	190	0.1
dfg	65	163	0.3
cmul	43	59	0.1
fft	274	1504	3.4
cmul in [11]	141	44	0.7
cmul†	137	381	2.3
dfg†	332	1465	105

† Simultaneous scheduling

Table 4. Code Generation And Address Optimization

Ex	TI 'C' Compiler O	A	Cyc	Optimization O	A	Cyc	Times Impr
hp1	48	39	87	22	6	28	3.10
lms	43	36	79	26	6	32	2.46
fft	108	87	195	99	7	106	1.83

7. Discussions and Conclusions

Although the model is solved as an ILP, the majority of examples in this paper could be solved using only one LP solution. This is not surprising since related research has shown that the use of logical inferences and horn clauses in ILP's has this characteristic[7]. Although the strict use of only horn clauses may require large number of constraints we found that the number of constraints generated in this model were manageable. An ILP approach was used instead of a production system since it is independent of the order of rules (therefore does not restrict formulation or ordering of rules), and it is easier to model complex constraints, including non-horn clauses. The optimality of the code generated by this modeling approach is only as good as the model itself. In other words if the constraints and arc mappings defined in the model are not complete, then the code generated may not be optimal.

Any scheduler such as one to minimize maximum density of variable lifetimes or the scheduler in [4] can be used in conjunction with this instruction selection, compaction and register allocation model. Other DSP processors can be modeled, such as the Motorola M56K and TMS320C3x. In combination with our addressing technique the savings in performance and codes size is even more significant, see table 4. For example in the *cmul* case the acutal final code using indirect addressing provides improvement over what was reported as the theoretical lower bound in [11].

In summary code size and performance improvements of 1.09 to 2.18 times (see table 2) were attained over previous research and commercial compilers. The technique presented in this paper performs simultaneous instruction selection, compaction and register allocation in very fast cpu times unlike other researched models[5]. Simultaneous scheduling and instruction selection, compaction, and register allocation can also be done with this model, however cpu times do increase (see table 3). In contrast to previous research which examined code compaction without instruction selection and register allocation [8], or code generation without compaction [1] the model presented in this paper can perform these tasks simultaneously. This research is important for industry since code generation for DSP processors or cores that minimizes code size, energy dissipation and maximizes performance, is critical to ensuring that the final product will be reliable, cost effective, and competitive. This research is supported in part by grants from NSERC and ITRC.

References

[1] G.Araujo, S.Malik, "Optimal Code Generation for Embdded memory Non-homogeneous register architectures", 8th International Symp on System Synthesis, 1995, p36-41.

[2] G.Araujo, S.Malik, M.T-C.Lee "Using Register-Transfer Paths in Code Generation of Heterogeneous Memory-Register Architectures", Design Automation Conference, 1996.

[3] A. Timmer,M.Strik, J.vanMeerbergen, J.Jess "Conflict Modeling and Instruction Scheduling in Code Generaton for In-house DSP Cores", Design Automation Conference, 1995.

[4] S.Liao, S.Devedas, K.Keutzer, S.Tjiang, A.Wang "Code Optimization Techniques for Embedded DSP Microprocessors" Design Automation Conference 1995.

[5] T.Wilson, G.Grewal, B.Halley, D.Banerji "An Integrated Approach to Retargetable Code Generation " 7th Int'l Symp High Level Synthesis, 1994 p70-75.

[6] R.Raman, I.Grossman,"Relation Between MILP Modeling

and Logical Inference for Chemical Process Synthesis" Dept of Chemical Eng, Technical Report, CMU, Jan 1990.

[7] Hooker, J.N. "Resolution vs. Cutting Plane Solution of Inference Problems: Some Computational Experience", Operations Research Letters, vol7,No1(1) 1988.

[8] P.Marwedel, G.Goossens Eds. **Code Generation for Embedded Processors**, Kluwer Academic Pub, 1995.

[9] Garey, Johnson **Computers and Intractability**, Freeman, 1979.

[10] E.Lee, "Programmable DSP Architectures: Part I and II", IEEE ASSP Magazine, Oct and Nov 1988.

[11] R.Leupers, P.Marwedel, "Time-Constrained Code compaction for DSP's", Trans on VLSI Systems, Vol.5,no.1, March 1997.

[12] C.Gebotys, "DSP Address Optimization Using A Minimum Cost Circulation Technique", to appear Int'l Conf on Computer-Aided Design, Nov 1997.

[13] S.Liao,S.Devadas,K.Keutzer,S.Tjiang,A.Wang, "Storage Assignment to Decrease Code Size", ACM SIGPLAN Conference on Programming Language Design and Implementation (PLDI), 1995.

[14] A.Brooke, D.Kendrick, A.Meeraus, **GAMS A User's Guide**, Scientific Press, 1992.

[15] **TMS320C2x User's Guide,** Texas Instrument Inc, 1996.

[16] P.E.Papamichalis, **Digital Signal Processing Applications with the TMS320 Family,** Vol.3, Prentice Hall, 1990.

Session 3

Novel Compilation and Optimization Issues

Embedded System Synthesis by Timing Constraints Solving

Krzysztof Kuchcinski

Dept. of Computer and Information Science
Linköping University
Sweden

Abstract

This paper presents an approach to embedded system synthesis which minimizes a system cost while implementing given timing requirements. The embedded system is represented by a set of finite domain constraints defining different requirements on processes timing, system resources and interprocess communication. The assignment of processes to processors and interprocess communications to buses as well as their scheduling are then defined as an optimization problem. A prototype system, based on constraint solving techniques, has been implemented in CHIP 5, the constraint logic programming system. Experimental results show that this approach can be efficiently used to define different system constraints and generate optimized system implementations.

1. Introduction

Embedded systems are digital systems which typically perform specific functions, such as process control, digital signal processing, image processing or multimedia applications. They are usually specified by communicating concurrent processes and implemented in software and hardware [7], [14]. The software is executed on a processor or several processors while hardware is usually provided as specialized ASICs. The communication channels are implemented by buses, point-to-point links or shared memories.

An important part of embedded system synthesis is partitioning of the system specification into processors, ASICs and communication channels as well as scheduling of different system functions. The resulted system implements a specified performance while providing a cost efficient system architecture [6], [9]. The system synthesis is usually based on the system structure, the timing constraint specifications and the execution profile of the system.

Different approaches to embedded system synthesis are proposed. In [9] authors define a flow graph model of the system and operation-level timing constraints. Operation delay and execution rate constraints capture timing properties of the system. They are used to develop the notion of satisfiability of constraints by a given implementation. Constraint satisfiability analysis is then used to define hardware and software partitions of functionality. In [5] a process graph, which represents a structure of processes and their interconnections, is defined. A node and an edge of the process graph have assigned weights which capture execution statistics. This information is used by the partitioning algorithm to make a decision whether a given process should be implemented in software or hardware. Two algorithms based on simulated annealing and tabu search have been implemented.

Integer Linear Programming (ILP) or Mixed Integer Linear Programming (MILP) based approaches are considered in [2], [12] and [13], for example. These methods produce optimal solutions while suffering usually from long execution times and practical limitation on the size of the problem. The MILP method requires a careful formalization of the problem as a set of linear inequalities. Usually, two types of variables are used; system parameter variables, such as timing variables, and binary variables. While the system variables represent real values assigned to system parameters, the binary variables are introduced to represent implementation decisions regarding the system configuration. The number of the binary variables and the related inequalities tends to be very large, even for moderate size problems, which results in long execution times for finding optimal solutions.

In this paper, a new method for embedded system partitioning and scheduling, is presented. The method is based on timing constraints solving techniques. A system is specified by a set of constraints over the finite domain of integers. The constraints are given as inequalities and specialized global constraints, and specify timing requirements, process precedence relations as well as resource constraints. The constraint solving technique is then used to find an optimal or suboptimal solution which satisfies given constraints and optimizes a given cost function. In our case, we use a branch and bound algorithm with

This work has been supported by the Wallenberg Foundation project "Information Technology for Autonomous Aircraft"

1080-1820/97 $10.00 © 1997 IEEE

depth-first-search or partial-depth-first-search method to find the system implementation. The prototype system has been implemented in CHIP 5 constrained logic programming system [4] and tested on several examples.

Constraint logic programming has already been used in the hardware design automation area. In [1] authors studied a use of Prolog with rational constraints, CLP(R), for symbolic timing verification. They modeled digital circuits as communicating processes and verified a large class of mixed synchronous and asynchronous specifications. The main idea of this work was to answer a question whether a design with specified delays could meet its timing constraints. By using constrained logic programming, the authors could handle not only constant component delays but component delays specified as range constraints. They could also determine ranges for delays that would still meet the constraints. In [8] authors developed a simple yet complete method for interface timing verification based on constraint logic programming and relational interval arithmetic in CLP(BNR) Prolog. They verify consistency, causality and compatibility of timing constraints.

Constraint Logic Programming system ECLIPSE has been used in [3] for retargetable self-test program generation. The authors generate first a code for a processor which is a set of partially ordered instructions. These instructions are then scheduled. Concurrently with scheduling the code compaction and binding is performed. This process, called global scheduling by the authors, is done by solving linear constraints over integer domain defined as a system of linear equations and inequalities.

Our approach uses constraint logic programming as well but differs from the described approaches in several ways. First, we make an extensive use of finite domain constraints without limiting them to linear equations and inequalities. The rational constraints are not used in our approach. This decision seems to be more natural for digital systems design. Second, we combine timing constraints and resource constraints and propose a new constraint representation for allocation and scheduling. Finally, we use constraints for synthesis by optimizing designs with imposed constraints. The optimization minimizes an explicitly given cost function while satisfying constraints set.

The rest of the paper is organized as follows. Section 2 introduces the system model based on the process graph which is later formalized in terms of finite domain constraints in section 3. The synthesis method, which includes both partitioning and scheduling, based on the proposed representation is introduced in section 4. Finally, the experimental results are presented in section 5 and the conclusions in section 6.

2. System Model

Consider the system specification consisting of six processes depicted as the process graph in Fig.1. Each node in

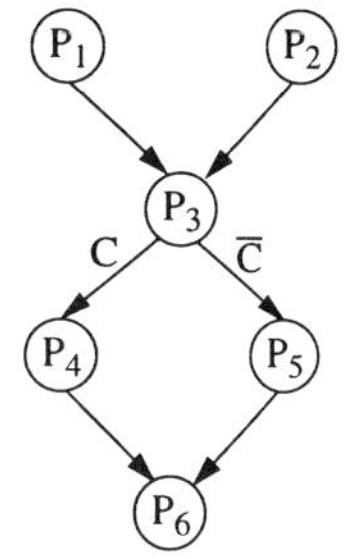

Fig. 1: An example of a process graph

this graph, (P_1, P_2, P_3, P_4, P_5, P_6), corresponds to a process and an edge connects two nodes if and only if there exist at least one direct communication channel between the corresponding processes. The process activates communication to another process, at the end of its execution, if there is an edge in the process graph connecting a sender process node with the receiver process node. For example, the process P_1 communicates with the process P_3 since there is the edge between nodes P_1 and P_3.

An edge with an assigned condition is a conditional edge (the edge connecting P_3 and P_4 in Fig. 1, for example) and a node with output conditional edges is a fork node (the node P_3 in Fig. 1). The fork node activates communication on those conditional edges whose assigned conditions are true. Conditional paths activated by the fork node meet at a joint node (the node P_6 in Fig. 1). The joint node is activated if the communication at all of the activated conditional paths took place.

In general, a process, represented by a node, is activated when a communication on all its active input channels, modeled by edges, took place. In the case of ordinary processes, the communication must take place on all its input channels while in the case of processes represented by joint nodes the communication has to take place only on the channels activated by the previously selected conditional paths.

The process graph is acyclic but an implicit loop between output nodes and input nodes is assumed. In the process graph, depicted in Fig.1, for example, after execution of the process P_6, the processes P_1 and P_2 are activated again. The execution can thus be repeated infinitely.

Each process has a deterministic execution time. This means that data-dependent loops are not allowed in a process body. A communication time is deterministic but can be different for different communication devices. In particular, we assume that the communication between two processes assigned to the same resource, either a processor or an ASIC, does not require any time.

Communication and synchronization between processes is allowed only at the beginning or at the end of the process execution, as indicated by the process graph. Many embedded systems fit into this model. For example, DSP program specifications based on synchronous data-flow model have these properties. If this restriction is too strong and it can not be fulfilled by the system specification, the specification has to be partitioned into a number of smaller processes. After this pre-partitioning communication channels need to be established to provide communication facilities between newly created processes.

The goal of the embedded system synthesis is to find an assignment of processes into processors/ASICs and communication channels into communication resources (e.g., buses, point-to-point links) which fulfils given timing constraints and minimizes system resources.

3. Finite Domain Constraints Model

The process graph is modeled as a set of finite domain constraints imposed on timing of processes and system resources. Since the constraints are defined for domain variables, we define first the variables which represent basic parameters of processes and resources and then introduce the basic constraints on these variables.

Definition 1. *Process.* A process is a 3-tuple $P=(\tau, \delta, \rho)$, where τ is an activation time of the process, δ is an execution time of the process, and ρ is a resource used to execute the process; τ, δ, ρ are the finite domain variables.

For example, for the process graph depicted in Fig. 1 the following definition of the process P_1 can be made: $P_1=(\tau_1, \delta_1, \rho_1)$, `τ1::0..50, δ1::[2, 5], ρ1::3..4`. The process P_1 is activated at the time between 0 and 50, its execution time is either 2 or 5 time units and it is executed on the resource 3 or 4. For a particular implementation, the final decision on allocating of processes to given processors/ASICs is made and unique values from domains of related domain variables are assigned to the parameters τ, δ, and ρ. For example, P_1=(0, 2, 4).

A process, represented by a 3-tuple, can be interpreted as a rectangle in the time/resource space having the following coordinates $((\tau_i, \rho_i), (\tau_i, \rho_i+1), (\tau_i+\delta_i, \rho_i), (\tau_i+\delta_i, \rho_i+1))$.

A single process specification does not include process graph information on process execution dependencies and possible system timing constraints. They are defined as the process graph constraints.

Definition 2. *Process graph.* If a process P_i is activated before a process P_j in the process graph the following inequality constraint is defined:

$$\tau_i+\delta_i \leq \tau_j$$

It can be noted that the specification of the process graph constraints can be used to capture duration and deadline constraints in specifying real-time systems and all possible binary timing relationships between processes [9]. The binary relations of the form "process P_i meets process P_j" or "process P_i overlaps process P_j" are easy to impose using basic structure of the process graph constraints. For example, the constraint $\tau_i+\delta_i = \tau_j$ specifies the relation "process P_i meets process P_j".

Definition 3. *Resource constraint.* If processes P_i and P_j are not conditional processes then the following constraint is defined:

$$\tau_i+\delta_i \leq \tau_j \lor \tau_j+\delta_j \leq \tau_i \lor \rho_i \neq \rho_j$$

The constraint imposes a requirement on two processes P_i and P_j that they can not be assigned to the same resource if at least a part of their executions overlaps. This constraint does not apply to conditional processes which are never executed at the same time and can share the same resource.

The above defined resource constraint has to be imposed on all processes which can be executed in parallel and thus their executions can overlap. It can lead to creation of many resource constraints. In the prototype, system defined in CHIP 5 constrained logic programming system, we a global constraint `diffn/1` [4] which makes use of a rectangle interpretation of processes, mentioned above. The `diffn/1` constraint takes as an argument a list of n-dimensional rectangles and assures that for each pair of *i*, *j* ($i \neq j$) of n-dimensional rectangles, there exist at least one dimension *k* where *i* is after *j* or *j* is after *i*. The n-dimensional rectangle is defined by a tuple $[O_1, ..., O_n, L_1, ..., L_n]$, where O_i and L_i are respectively called the origin and the length of the n-dimensional rectangle in *i-th* dimension.

Definition 4. *Rectangle based resource constraint.* If processes P_i and P_j are not conditional processes then the constraint on their respective rectangles, $R_i=[\tau_i, \rho_i, \delta_i, 1]$ and $R_j=[\tau_j, \rho_j, \delta_j, 1]$, imposes that rectangles R_i and R_j can not overlap.

The rectangle based resource constraints can be easily extended to handle conditional processes. The presented formulation of the resource constraints uses 2-dimensional rectangles in the time/resource space. The main idea of representing conditional processes is to extend rectangles into higher dimensions. In principle, one more dimension for every new condition is used. Thus the representation of the process graph depicted in Fig. 1 requires 3-dimensional rectangles.

A process execution time is deterministic but it is not constant for all resources. Microprocessors usually require longer execution time for the same process than ASICs. Different microprocessors provide different execution time for the same process depending on the processor's clock frequency and its architecture. The model makes it possible to define the process execution time as a domain variable capturing several execution time values. In addition, the mapping which binds a given execution time to a given resource is defined. This is achieved by defining additional constraints on execution time. Basically, it assigns a specific execution time for a process depending on the selected resource. In constraint logic programming, it is done by the `element/3` constraint. It assigns one value from the list of values to the third argument. The value is selected based on the first argument which works as an index.

Definition 5. *Execution time constraint.* An execution time, δ_i, for a process P_i on a resource ρ_i is defined by the following constraint:

$$\texttt{element}(\rho_i,\ [T_{i1},\ T_{i2},\ \ldots,\ T_{iN}],\ \delta_i).$$

Example 1. Consider the process graph depicted in Fig. 1. The following constraints are created to represent the process graph. An index of τ, ρ and δ represents a number of a process while indices *0* and *e* represent the start and the end time of the computation respectively.

$\tau_1::0..50,\ \delta_1::[1,\ 2],\ \rho_1::1..2,$
$\tau_2::0..50,\ \delta_2::[2,\ 3],\ \rho_2::1..2,$
$\tau_3::0..50,\ \delta_3::[4,\ 6],\ \rho_3::1..2,$
$\tau_4::0..50,\ \delta_4::[5,\ 3],\ \rho_4::1..2,$
$\tau_5::0..50,\ \delta_5::[2,\ 3],\ \rho_5::1..2,$
$\tau_6::0..50,\ \delta_6::[3,\ 1],\ \rho_6::1..2,$

$\tau_0 \le \tau_1,\quad \tau_0 \le \tau_2,\quad \tau_4 + \delta_4 \le \tau_6,\ \tau_5 + \delta_5 \le \tau_6,$
$\tau_1+\delta_1 \le \tau_3,\quad \tau_2+\delta_2 \le \tau_3,\quad \tau_3+\delta_3 \le \tau_4,\quad \tau_3+\delta_3 \le \tau_5,$
$\tau_6+\delta_6 \le \tau_e,$

$\texttt{C::1..2, NotC::1..2, C} \neq \texttt{NotC},$
$\texttt{diffn}([[\tau_1,\rho_1,1,\delta_1,1,2],\ [\tau_2,\rho_2,1,\delta_2,1,2],$
$[\tau_3,\rho_3,1,\delta_3,1,2],\ [\tau_4,\rho_4,\texttt{C},\delta_4,1,1],$
$[\tau_5,\rho_5,\texttt{NotC},\delta_5,1,1],\ [\tau_6,\rho_6,1,\delta_6,1,2]]),$

$\texttt{element}(\rho_1,[1,2],\delta_1),\quad \texttt{element}(\rho_2,[2,3],\delta_2),$
$\texttt{element}(\rho_3,[4,6],\delta_3),\quad \texttt{element}(\rho_4,[3,5],\delta_4),$
$\texttt{element}(\rho_5,[2,3],\delta_5),\quad \texttt{element}(\rho_6,[1,3],\delta_6).$

The argument of `diffn/1` constraint is a list of 3-dimensional rectangles. As it has been indicated, the third dimension represents the conditional execution of processes depending on the condition `C` or `NotC`. All the unconditional processes (i.e., P_1, P_2, P_3 and P_6) have this coordinate 1 and the height 2. The conditional processes start at the third coordinate either 1 or 2, depending on the condition, and have a height 1. They can share the same time/resource place since they can be placed "one on top of the other". Other processes can not be placed together with conditional ones since they have a height 2. The `element/3` constraint defines mappings between the process execution time and the selected resource. In the example, we use only two resources which represent different execution times for all processes.

Constraints specified for a given design are an input to a constraint solver, in our case CHIP Prolog system, and are automatically verified for consistency. Domains of variables are adjusted based on existing constraints. For example, the constraint solver adjusts $\tau_1::0..40$, $\tau_2::0..39$, $\tau_3::2..41$, $\tau_4::6..45$, $\tau_5::6..46$ and $\tau_6::10..49$. It directly prunes the search space and creates basis for implementation of heuristic partitioning and scheduling algorithms.

Communication between processes is specified for all interacting processes and is taken into account during synthesis. Conceptually, the communication between two processes can be considered as another process which uses a specialized communication resource, a bus for example. The communication process is assigned to all edges connecting two processes. The model is similar to the previously introduced model of processes based on inequalities and rectangles. The communication time between processes is incorporated into process graph constraints which is achieved by replacing the inequality defined in the definition 2 by another two inequalities given below.

Definition 6. *Process graph with communication.* If a process P_i starts communication with a process P_j at $\tau_{i,j}$ and continues this communication for $\delta_{i,j}$ time units, the following inequality constraints are defined:

$$\tau_i+\delta_i \le \tau_{i,j},\quad \tau_{i,j} + \delta_{i,j} \le \tau_j$$

The communication time between two arbitrary processes depends on a communication resource selection as well as on an assignment of processes to processors and ASICs. While the communication time of the communication resource can be constrained by the execution time constraint given in the definition 5, the communication time depending on the processes assignment is different. If two processes are assigned to the same resource the communication overhead is very small. Assignment of processes to separate resources introduces additional overhead since additional communication components, such as a shared bus or a direct point-to-point link, have to be used.

In this paper, an assumption is made that processes assigned to the same processor/ASIC have a communication time equal zero while in the other cases the communication time depends on communication resource characteristics. The modeling of this situation requires conditional propagation of constraints. The decision on constraint satisfaction is delayed until a condition is true or false for all possible values of the domain variables. Syntactically, it is specified using `if-then-else` constraints. The following definition introduces communication time constraint.

Definition 7. *Communication time.* If a process P_i communicates with a process P_j then the following conditional constraint is defined:

$$\textbf{if } \rho_i = \rho_j \textbf{ then } \delta_{i,j} = 0 \textbf{ else } \texttt{element}(\rho_{i,j}, [T_{ij1}, T_{ij2}, \ldots T_{ijN}], \delta_{i,j}).$$

Example 2. Consider the constraints defined in the example 1. In this example, we change process graph constraints to process graph constraints with communication and extend them with the communication constraints given in the definition 7. We use a single communication resource which has a single communication time for different communications between processes and thus we do not need to use `element/3` constraint.

$$\tau_0 \le \tau_1,\quad \tau_0 \le \tau_2,\quad \tau_6 + \delta_6 \le \tau_e,$$
$$\tau_1+\delta_1 \le \tau_{13},\quad \tau_{13} + \delta_{13} \le \tau_3,$$
$$\tau_2+\delta_2 \le \tau_{23},\quad \tau_{23} + \delta_{23} \le \tau_3,$$
$$\tau_3+\delta_3 \le \tau_{34},\quad \tau_{34} + \delta_{34} \le \tau_4,$$
$$\tau_3+\delta_3 \le \tau_{35},\quad t_{35} + \delta_{35} \le \tau_5,$$
$$\tau_4+\delta_4 \le \tau_{46},\quad t_{46} + \delta_{46} \le \tau_6,$$
$$\tau_5+\delta_5 \le \tau_{56},\quad t_{56} + \delta_{56} \le \tau_6,$$

$$\delta_{13}::[0,1],\quad \delta_{23}::[0,2],\quad \delta_{34}::[0,2],$$
$$\delta_{35}::[0,2],\quad \delta_{46}::[0,1],\quad \delta_{56}::[0,1],$$
$$\rho_{13}::1..1,\quad \rho_{23}::1..1,\quad \rho_{34}::1..1,$$
$$\rho_{35}::1..1,\quad \rho_{46}::1..1,\quad \rho_{56}::1..1,$$

$$\textbf{if } \rho_1 = \rho_3 \textbf{ then } \delta_{13} = 0 \textbf{ else } \delta_{13} = 1,$$
$$\textbf{if } \rho_2 = \rho_3 \textbf{ then } \delta_{23} = 0 \textbf{ else } \delta_{23} = 2,$$
$$\textbf{if } \rho_3 = \rho_4 \textbf{ then } \delta_{34} = 0 \textbf{ else } \delta_{34} = 2,$$
$$\textbf{if } \rho_3 = \rho_5 \textbf{ then } \delta_{35} = 0 \textbf{ else } \delta_{35} = 2,$$
$$\textbf{if } \rho_4 = \rho_6 \textbf{ then } \delta_{46} = 0 \textbf{ else } \delta_{46} = 1,$$
$$\textbf{if } \rho_5 = \rho_6 \textbf{ then } \delta_{56} = 0 \textbf{ else } \delta_{56} = 2,$$

An efficient way of improving performance of embedded systems is to pipeline their subsequent computations. The system performance is improved while the related increase in the system cost is usually not very high. Pipelining is used very often in DSP applications or microprocessors, for example. Representing pipelining using our modeling technique is relatively easy. It requires an extension of the resource constraints of the process graph represented by n-dimensional rectangles (definition 4), as defined below.

Definition 8. *Resource constraints for pipelined computations.* If a process graph, without fork nodes, is to be pipelined with the initiation rate k time units and n pipeline stages than the following resource constraints are defined:

$$\tau_{1,k} = k + \tau_1, \quad \tau_{n,k} = k + \tau_n,$$
$$\ldots$$
$$\tau_{1,P} = (n-1)*k + \tau_1, \quad \tau_{n,P} = (n-1)*k + \tau_n,$$

$$\texttt{diffn}(\ [[\tau_1,\rho_1,\delta_1,1],\ldots,\ [\tau_n,\rho_n,\delta_n,1],$$
$$[\tau_{1,k},\rho_1,\delta_1,1],\ldots,\ [\tau_{n,k},\rho_n,\delta_n,1],$$
$$\ldots,$$
$$[\tau_{1,P},\rho_1,\delta_1,1],\ldots,\ [\tau_{n,P},\rho_n,\delta_n,1]]).$$

Intuition behind this formalization is straightforward. Introducing a pipeline of the initiation rate *k* time units is equivalent to a placement of *k* copies of rectangles starting at positions *k*, *2*k*, *3*k*, etc. This prevents a synthesis algorithm to place operations at forbidden locations, which will be used by subsequent pipeline computations. There exist a clear relation to reservation tables (see, for example [10]) know from computer architecture.

The definition 8 defines basic resource conflict constraints for unconditional processes. When a computation contains conditional processes this constraint is too weak. It does not cover the situation when two conditional processes, such as processes P_4 and P_5 in the Fig. 1, which are run in two subsequent computations want to share the same resource. In this case, additional constraints has to be added to disallow this situation. They can, for example, be formulated as additional `diffn/1` constraints.

Example 3. Consider our example depicted in Fig. 1 with the constraints defined in the examples 1 and 2. In this example, we will define constraints for 2 stage pipeline with initiation rate k.

$$\tau_{1,1} = k + \tau_1,\quad \tau_{2,1} = k + \tau_2,\quad \tau_{3,1} = k + \tau_3,$$
$$\tau_{4,1} = k + \tau_4,\quad \tau_{5,1} = k + \tau_5,\quad \tau_{6,1} = k + \tau_6,$$

$$\texttt{diffn}([[\tau_1,\rho_1,1,\delta_1,1,2],\ [\tau_2,\rho_2,1,\delta_2,1,2],$$
$$[\tau_3,\rho_3,1,\delta_3,1,2],\ [\tau_4,\rho_4,\texttt{C},\delta_4,1,1],$$
$$[\tau_5,\rho_5,\texttt{NotC},\delta_5,1,1],\ [\tau_6,\rho_6,1,\delta_6,1,2],$$
$$[\tau_{1,1},\rho_1,1,\delta_1,1,2],\ [\tau_{2,1},\rho_2,1,\delta_2,1,2],$$
$$[\tau_{3,1},\rho_3,1,\delta_3,1,2],\ [\tau_{4,1},\rho_4,\texttt{C},\delta_4,1,1],$$
$$[\tau_{5,1},\rho_5,\texttt{NotC},\delta_5,1,1],\ [\tau_{6,1},\rho_6,1,\delta_6,1,2]]),$$

$$\texttt{diffn}([[\tau_4,\rho_4,\delta_4,1],\ [\tau_{5,1},\rho_5,\delta_5,1]]),$$
$$\texttt{diffn}([[\tau_{4,1},\rho_4,\delta_4,1],\ [\tau_5,\rho_5,\delta_5,1]]).$$

Once all constraints are defined they can be checked for consistency. It is done automatically by the Prolog system which either succeeds or fails to create constraints. If the system fails the constraints are not consistent. If it succeeds, more analysis is still needed because the final checking of some constraints have been delayed.

4. System Synthesis

The embedded system, defined as a set of constraints on processes, communication and resources, has to be partitioned into processors, ASICs and communication channels. The processes have to be scheduled on the selected resources. The main goal is to implement a specified system performance while providing a cost efficient system architecture. This is a typical optimization problem which can be defined in terms of our constraints.

The system specification can also be used to make a quick evaluation of different system implementations and their related cost as well as performance. In this case, the constraints provide a way to automatically "guard" system requirements while constructing new system architectures.

The system synthesis is solved as an optimization problem on constrained domain variables defined for a given system, such as τ_i, and ρ_i. Each domain variable has a finite domain of values. A value, from this domain, can be assigned to the variable if it does not contradict with defined system constraints. The optimization is defined as an assignment of values to all variables which minimizes/maximizes a given cost function. For the synthesis problem, two cases have been identified. First, the number of resources is minimized while a system execution time is given. Second, the resources of the system are given and the execution time is minimized.

Different optimization algorithms can be used to select an assignment of values to domain variables. In this paper, the branch and bound algorithm with the depth-first-search (dfs) method is used. The algorithm assigns values to pairs of domain variables (τ_i, ρ_i) using different heuristics to select an order of this assignment. The algorithm finds an optimal solution for the synthesis problem, however, it is *NP*-complete and in some cases its execution time can be very large. In these cases heuristic optimization methods, such as a partial-depth-first-search (pdfs) can be used instead of dfs. Pdfs reduces the complexity of the search but does not provide a guarantee to find an optimal solution.

Example 4. Consider the system definition given in the example 1. Fig. 2 shows results of two different optimizations of the same system. The optimization of resources with the execution time limit of 15 time units gives the assignment of all processes to one processor while optimization of time with the resource limit of two processors finds the processor assignment which has 13 execution steps. In both cases, processes P_4 and P_5 are assigned to the same processors since their execution is conditional and they will never be executed simultaneously.

5. Experimental Results

The experiments have been carried out using a prototype implementation of the synthesis system implemented in CHIP 5, the constrained logic programming system [4]. This is a Prolog system with constraints solvers over boolean, finite and rational domains. In the experiments, the finite domain solver has been used only. We have used a branch-and-bound optimization algorithm with two search methods: depth-first-search (dfs) and partial-depth-first-search (pdfs). While dfs is *NP*-complete, the pdfs has, in these experiments, the complexity of $O(N^2)$ or $O(c*N)$, where N is the number of elements which needs to be assigned. In our case, N is the number of processes represented as pairs (τ_i, ρ_i). Selection of the search methods is indicated in the table 1. The experiments presented in the table 2 use pdfs. Constraints have been imposed using previously defined methods. In the tables, we give the number of constraints for each experiment and the execution time on the SPARC workstation to obtain either an optimal (dfs) or suboptimal solution (pdfs). The table 1 gives also the number of nodes visited by the branch and bound algorithm (denoted as "B&B Nodes").

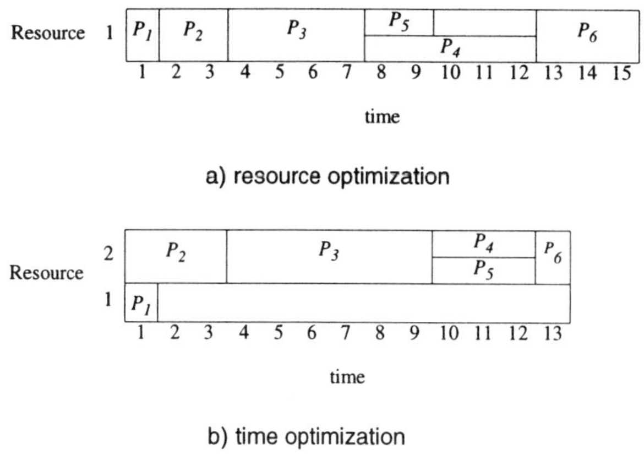

Fig. 2: Synthesis alternatives for the example 1

We have examined an example of a system consisting of 17 processes depicted in Fig. 3 as a process graph. For the experiments the following processes' execution times have been assumed δ_1::[2,3], δ_2::[2,4], δ_3::[8,12], δ_4::[3,5], δ_5::[2,3], δ_6::[3,5], δ_7::[2,3], δ_8::[2,4], δ_9::[3,5], δ_{10}::[3,5], δ_{11}::[4,6], δ_{12}::[4,6], δ_{13}::[5,8], δ_{14}::[1,2], δ_{15}::[4,6], δ_{16}::[2,4], δ_{17}::[1,2], where for each pair of values

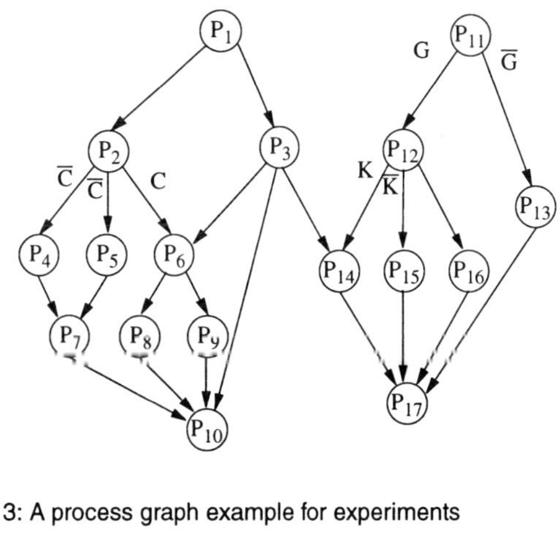

Fig. 3: A process graph example for experiments

the first one represents the execution time on the processor type 1 and the second one the execution time on the processor type 2. The communication time for processes assigned to the same processor is 0 and for processes communicating through a bus are the following: $\delta_{1,2}=1$, $\delta_{1,3}=1$, $\delta_{2,4}=0$, $\delta_{2,5}=3$, $\delta_{2,6}=0$, $\delta_{3,6}=2$, $\delta_{3,10}=2$, $\delta_{3,14}=2$, $\delta_{4,7}=3$, $\delta_{5,7}=1$, $\delta_{6,8}=3$, $\delta_{6,9}=3$, $\delta_{7,10}=2$, $\delta_{8,10}=2$, $\delta_{9,10}=2$, $\delta_{2,5}=3$, $\delta_{11,12}=1$, $\delta_{11,13}=2$, $\delta_{12,14}=1$, $\delta_{12,15}=3$, $\delta_{13,16}=1$, $\delta_{13,17}=2$, $\delta_{14,17}=2$, $\delta_{15,17}=1$, $\delta_{16,17}=2$.

Several experiments have been carried out to check the feasibility of the constrained based synthesis approach. The successive experiments introduce new constraints and more complexity into the optimization problems. The results are presented in Table 1.

The experiment 1 assumes a given assignment of processes to processors and the goal is to find a minimal number of execution steps for the system by finding the schedule for processes and communications on the bus. The following assignment is assumed for processor 1 (P_1, P_2, P_4, P_6, P_9, P_{10}, P_{13}), processor 2 (P_3, P_5, P_7, P_{11}, P_{14}, P_{15}, P_{17}) and processor 3 (P_8, P_{12}, P_{16}). All three processors are of type 2 (p2). In the experiment 2, the system finds a better assignment of processes to two processors p2 and the related communication assignment to two buses is found in the experiment 3. Finally, both the processes and communication assignment for this configuration is found in the experiment 4. In experiments 5, 6 and 7, we assume that two types of processors are available, p1 and p2. The processor assignment is found in the experiment 5 and buses for this assignment are allocated in the experiment 6. In the experiment 7 both processors and buses are optimized. Two

Table 1: Results of system synthesis for the example in Fig. 3

Experiment	Constraints	dfs/ pdfs	Execution steps	Components	CPU time (s)	B&B Nodes
(1) pre-assigned	26 ineq. 37 5-D rect.	dfs	39	3 p2	0.23	62
(2) p2 assignment	26 ineq. 23 cond. 17 5-D rect.	dfs	33	2 p2	0.42	56
(3) bus assignment for (2)	26 ineq. 23 cond. 30 5-D rect.	dfs	33	2 p2, 2 buses	0.09	42
(4) p2 & bus assignment	26 ineq. 23 cond. 30 5-D rect.	dfs	33	2 p2, 2 buses	0.97	86
(5) p1 & p2 assignment, no bus	26 ineq. 23 cond. 17 5-D rect. 17 element	dfs	22 24	1 p1, 2 p2 1 p1, 1 p2	0.54 0.38	68 19
(6) bus assignment for (5)	26 ineq. 23 cond. 30 5-D rect. 17 element	dfs	22 24	1 p1, 2 p2, 2 buses 1 p1, 1p2, 1 bus	0.04 0.02	42 42
(7) p1, p2 & bus assignment	26 ineq. 23 cond. 30 5-D rect. 17 element	dfs	22 24	1 p1, 2 p2, 2 buses 1 p1, 1 p2, 1 bus	2.24 0.63	139 606
		pdfs	23 25	1 p1, 2 p2, 2 buses 1 p1, 1 p2, 1 bus	8.66 9.76	435 606

Table 2: Results of pipelined system synthesis for the example in Fig. 3

Experiment	Constraints	Execution steps/stage	Components	CPU time (s)
(1) p2-type processors	26 ineq. 74 5-D rect. 28 2-D rect.	18	4 p2, 2 buses	155.57
		21	3 p2, 2 buses	454.63
		22	3 p2, 1 bus	30.28
		31	2 p2, 1 bus	275.18
(2) p1-type processors	26 ineq. 74 5-D rect. 28 2-D rect.	12	4 p1, 2 buses	218.84
		14	3 p1, 2 buses	199.46
		18	3 p1, 1 bus	183.94
		19	2 p1, 1 bus	173.95
(3) 1 p1-type processor and p2-type processors	26 ineq. 74 5-D rect. 28 2-D rect. 17 element	14	1 p1, 4 p2, 3 buses	416.73
		16	1 p1, 3 p2, 2 buses	426.83
		18	1 p1, 2 p2, 2 buses	345.30
		20	1 p1, 2 p2, 1 buses	531.32

different solutions are found, depending on the resource constraints. Both dfs and pdfs are used. While dfs guarantees to find an optimal solution pdfs finds a solution without guarantee that it is optimal. For example, in the experiment 7, in both cases, pdfs finds the solutions which requires one more execution step than the optimal solution found by dfs.

The presented experiments provide guidelines on the efficient use of the constraints solving techniques for system synthesis. While the experiment 7 includes all constraints, it could be too time consuming for synthesis of large systems. In this case, the combination of methods used in the experiment 5 and 6 can be used. The method performs first the processes assignment and scheduling and then allocates buses and schedules communication transactions.

Table 2 shows the pipeline synthesis results of the process graph which has been used in the previous experiments. The execution time data are the same as defined previously. A two stage pipeline has been defined using `diffn/1` constraint together with additional resource constraints on conditional processes executed in subsequent computations.

The results have been generated for different implementation alternatives using processors of type one and two. The pdfs with complexity $O(c*N)$ has been used in the experiments. The optimization strategy minimizes resources (processors and busses) under the stage length limit. In all cases, the pipeline stage execution time is shorter than the sequential executions time presented in Table 1.

6. Conclusions

In this paper, a new method for system synthesis is presented. The method is based on constraint solving techniques and has been implemented in the constraint logic programming system CHIP 5. Its usability has been demonstrated by applying it to several examples representing different problem complexity.

The developed method provides several advantages comparing to graph oriented synthesis methods. It introduces a clear distinction between formalization of system constraints and its synthesis by optimization of selected parameters. For the same formalization, different optimization methods can be used. In this paper, we have concentrated on depth-first-search branch and bound algorithms but other specific synthesis oriented heuristic algorithms can be considered as well as already known methods can be applied.

MILP and ILP based techniques for system synthesis use similar formulation for timing constraints but they need to extend their formulation with binary decision variables. In our method, this is not necessary since conditional constraints and specialized constraints, such as `element/3`, solves these kind of problems. It makes the problem formulation much easier and more readable. It is very important when defining constraints for systems containing many heterogeneous constraints which are difficult to represent using binary variables.

A solution of a set of constraints and their optimization is, in principle, an enumeration problem with exponential complexity. A search space can, however, be pruned in several ways. First, for tightly constrained problems constraints propagation dominates an execution and prunes the search space. The constraint propagation algorithms have polynomial complexity. Second, the branch and bound algorithm can cut away many search branches if the cost function provides a good estimation of the solution before all domain variables are instantiated. It is very often the case when constraints are used to define the cost function. Finally, sub-optimal solutions can be obtained using heuristics. All these arguments make constraint solving techniques an attractive alternative to MILP and ILP methods for system-level design problems.

New constraints can easily be defined and considered in the system synthesis. This is an important feature of this approach since system specification usually involves many heterogeneous constraints. A rate constraint, which is very common in real-time applications, or memory access requirements can be introduced as additional constraints. Information from a designer, who makes decisions and assigns selected values to domain variables, can also be introduced. It was illustrated in our experiments by using predefined assignment of processes to processors, for example.

In the formalization, presented in this paper, only acyclic graphs can be considered. It is usually a major limitation for realistic designs. An extension, which makes it possible to introduce hierarchy and iterative computations similar to [11] can be used in this case.

References

[1] T. Amon and G. Boriello, *An Approach to Symbolic Timing Verification*, 29th ACM/IEEE Design Automation Conference, pp. 410-413, 1992.

[2] A. Bender, *Design an Optimal Loosely Coupled Heterogeneous Multiprocessor System*, Proc. The European Design and test Conference, March 11-14, 1996, Paris, France, p. 275-281.

[3] U. Bierker and P. Marwedel, *Retargetable Self-Test Program Generation Using Constraint Logic Programming*, 32nd ACM/IEEE Design Automation Conference, pp. 605-611.

[4] *CHIP, System Documentation*, COSYTEC, 1996.

[5] P. Eles, Z. Peng, K. Kuchcinski and A. Doboli, *System Level Hardware/Software Partitioning Based on Simulated Annealing and Tabu Search*, Design Automation for Embedded Systems Journal, vol. 2, no. 1, January 1997.

[6] R. Ernst, J. Henkel, T. Benner, *Hardware-Software Co-Synthesis for Microcontrollers*, IEEE Design & Test of Computers, September 1993, pp. 64-75.

[7] D. D. Gajski, F. Vahid, *Specification and Design of Embedded Hardware-Software Systems*, IEEE Design & Test of Computers, Spring 1995, pp. 53-67.

[8] P. Girodias and E. Cerny, *Interface Timing Verification with Delay Correlation Using Constraint Logic Programming*, Proc. European Design and Test Conference, March 17-20, 1997, Paris, France.

[9] R.K. Gupta and G. de Micheli, *A Co-Synthesis Approach to Embedded System Design Automation*, Design Automation for Embedded Systems Journal, vol. 1, no. 1-2, January 1996.

[10] K. Hwang, *Advanced Computer Architecture: Parallelism, Scalability, Programmability*, McGraw-Hill, Inc., 1993.

[11] D. Ku and G. De Micheli, *Relative Scheduling under Timing Constraints: Algorithms for High-Level Synthesis of Digital Systems*, IEEE Transactions on Computer-Aided Design of Integrated Circuits and Systems, vol. 11, no. 6, June 1992.

[12] S. Prakash and A. Parker, *SOS: Synthesis of Application-Specific Heterogeneous Multiprocessor Systems*, Journal of Parallel and Distributed Computing, 16, 1992.

[13] S. Prakash and A. Parker, *Synthesis of Application-Specific Multiprocessor Systems Including Memory Components*, Journal of VLSI Signal Processing, 8, 1994.

[14] W. H. Wolf, *Hardware-Software Co-Design of Embedded Systems*, Proceedings of the IEEE, vol. 82, no. 7, July 1994, pp. 967-989.

Reducing the complexity of ILP formulations for synthesis

Anne Mignotte Olivier Peyran
Laboratoire de l'Informatique du Parallélisme
École Normale Supérieure de Lyon
46, Allée d'Italie, 69364 LYON Cedex 07, France
Anne.Mignotte@ens-lyon.fr, Olivier.Peyran@ens-lyon.fr

Abstract

Integer Linear Programming (ILP) is commonly used in high level and system level synthesis. It is an NP-Complete problem (in general cases). There exists some tools that give an optimal solution for small ILP formulations. Nevertheless, these tools may not give solutions for complex formulations. In this paper, we present a solution to overcome the problem of complexity in ILP formulations. We propose a partitioning methodology based on a constraint graph representing all the constraints included in any ILP formulation. To direct the partitioning, the constraint graph nodes are grouped to represent Data Flow Graph (DFG) nodes. This reduced constraint graph can be used to partition any ILP formulation based on DFG. We illustrate this method on ILP formulation for scheduling. Results on ILP scheduling formulations are promising.

1 Introduction

Integer linear programming (ILP) is a widely used technique in high level and system level synthesis. Common applications are resolution of NP-complete problems. For instance, [5] uses ILP for throughput and latency optimization when algorithm/architecture matching, retiming and pipelining are considered simultaneously. ILP is also used for DSPs code generation and embedded systems. For instance, [14] gives a solution to the problem of code compaction with real-time constraints for processors offering instruction-level parallelism; [21] presents an ILP-based code placement method for embedded software to maximize hit ratios of instructions caches. ILP is also widespread for HW/SW partitioning [2, 11, 17]. In the field of system level synthesis, one can also cite [20, 1], which deal with the optimization of heterogeneous multiprocessors systems. Another example is [19], whose method generates a static task execution schedule along with the structure of the multiprocessor system and a mapping of subtasks to processors. A part of the previous problem, scheduling, is a very common application of ILP: [10] presents different formulations for different usual scheduling problems; [8], [13] propose solutions to the more general problem of performing scheduling and resource allocation simultaneously; [3] describes a methodology to solve a scheduling problem in a 3-dimensional design space, including the usual area and schedule length dimensions plus the clock length dimension, using module libraries.

In [15], we consider the problem of performing scheduling and operator type selection simultaneously. Operands can be represented in various number systems (redundant or non redundant ones). This leads to architectures with different type of arithmetics (*mixed arithmetics*), and the insertion of conversion nodes between redundant output and non redundant input of operators is needed. As conversions are "real" operations (in fact additions with carry propagation), the operator type selection has to be done while scheduling. A heuristic have been proposed [16], but the results were not convincing. Thus, we explored the ILP approach. Our ILP formulation has been tested using *LP_ SOLVE* (see table 1). The results are optimum, and the computation times remain small for small examples.

Nevertheless, ILP is a NP-complete problem, and our ILP solver could not find any solution for reasonable size examples, because of numerical instability. Complexity is the major drawback of ILP. There are solutions to overcome this problem by directly modifying the formulation. In the next section, we present some of these techniques. However, they usually do not largely decrease the complexity, and are always specific to each problem. Thus, some partitioning techniques have been proposed in the literature. However they usually do not take into account all the ILP constraints (resource constraints and dependency constraints) as well as the ILP size. In section 3, we propose a general solution which partitions the problem into several small ILP formulations, separately solved taking all the constraints into account. Section 4 discusses the results and extensions of this method.

2 Shortening the ILP formulation

In this section, we present solutions that are specific to our resource constraint scheduling problem using mixed arithmetic. More precisely, all these solutions have been tested on the 5th order elliptic wave filter benchmark.

Our ILP formulation for scheduling is based on two main constraints: resource constraints, and data dependency constraints which are given by a DFG. The formulation is inspired by Hwang *et al.* one [10]. The variables and constants used are the following:

benchmarks	Nb equations	Nb variables	Nb operations	Nb cycles	CPU
2 way method	53	62	6	5	6.2 s
Fast Fourier Transform	51	78	7	4	3.6 s
Differential equation	99	132	11	5	17.6 s
3 way method	248	410	25	?	no sol.
3 way method classic	91	126	25	10	51.1 s
5th order elliptic wave filter (EWF)	383	991	34	?	no sol.
5th order EWF, $k = -1/0$	258	586	34	?	no sol.
5th order EWF, $k = \pm 1$, red. outputs	287	665	41	?	no sol.
5th order EWF classic	119	133	34	16	4 min 44.8 s

Table 1: ILP results using *LP_SOLVE*.

- ⋆ $x_{i,j} = 1$ if operation o_i is scheduled into cycle j; otherwise, $x_{i,j} = 0$.
- ⋆ T is the final number of cycles that we wish to minimize and N_t is the number of resources of type t.
- ⋆ s is an overestimation of T.
- ⋆ L_i (resp., S_i) is the latest (resp. earliest) possible time to schedule operation o_i. They are computed according the classical "As Late As Possible" and "As Soon As Possible"schedules considering that we have s cycles.

We have kept the same conventions and extended them to our specific problem: if o_i is a classical operation (addition, subtraction, multiplication...), it is also related to variable $x_{i,j}$, with $j \in [S_i, L_i]$. Our model inserts a virtual conversion after each operation. Therefore we need a new variable, $x_{k,j}$, representing the conversion. Our ILP formulation will not be detailed in this paper.

The complexity of an ILP formulation is the product of the number of equations by the number of variables. Concerning scheduling, these two values depend on the number of operations, n, and on the initial bound of the number of cycles, s (in our case, both grows in $O(s * n)$). The s value is a large overestimation, and considering the operation frames (difference between L_i and S_i) is more accurate.

To reduce the ILP formulation, the number of operations could obviously not be decreased. The only solution is to shorten the operation frames. It decreases the number of variables, and also the number of equations, as some resource constraints may disappear (if some juxtaposed operation frames become disconnected). Obviously, this does not guarantee optimality, as all the possible resulting schedules are not considered anymore. The problem is to find a good shortened frame (i.e an as small as possible frame containing the optimum schedule with an high probability).

A first method consists in producing an optimal linear programming solution using real variables. From this solution, we prune the ILP formulation by considering, for the operation possible schedules, the integer values that are close to the real-number solutions. However, a real-number solution is not an approximation of an integer solution. In some cases, this reduction can lead to a solution space which does not contain any integer solution. In opposite cases, the frame still remains too large to solve the instability problem.

Thus, we tried another technique: using an heuristic solution (for instance the list scheduling) to bound the frame of each operation, the frames are redefined as $T(i) \pm k$, where $T(i)$ is the scheduled cycle of operation o_i, and k is as small as possible. Nevertheless, the frames were still too large with $k = 1$, as shown table in 1. We even tried this method with frames equal to $\{T(i) - 1, T(i)\}$, and we did not obtain any solution either.

This solution is interesting when the initial operation frames are large. However, in such a case, reducing the frames will reduce the solution space, increasing the probability of "loosing" the optimal schedule. On the opposite, if operations have small mobility, the optimal schedule has an high probability of staying in the solution space, but the complexity can not be largely decreased, as time frames are already short. Therefore, we have explored another technique, which is addressed in the following sections, that consists in partitioning the problem in order to get several smaller ILP formulations, which are separately solved.

3 Partitioning

3.1 Partitioning methodology

The initial DFG is partitioned. Each partition can be considered as a separate design, and is scheduled using a separate ILP formulation. We obtain several optimal local schedules which are concatenated in order to obtain the final global schedule. The main difficulty is to find a partitioning algorithm, as there are two constraints to deal with: all the interdependencies between partitions and their size.

Some related methods of partitioning have already been proposed. For instance, Hwang experimented an approach [9], called "zone scheduling". It partitions the control steps into zones, and decides which operation will be scheduled into a zone and which one will be "delayed" into the next zone. Their model can be turned into an optimal ILP scheduling, a list scheduling, or one in between. However, their goal is more to find better solutions than those achieved by list scheduling rather than finding near optimal solutions when ILP does not succeeds. Depuydt et al. have a solution based on clustering techniques [4]. It does not take into account the resource constraints but variable time frames to reduce the register cost.

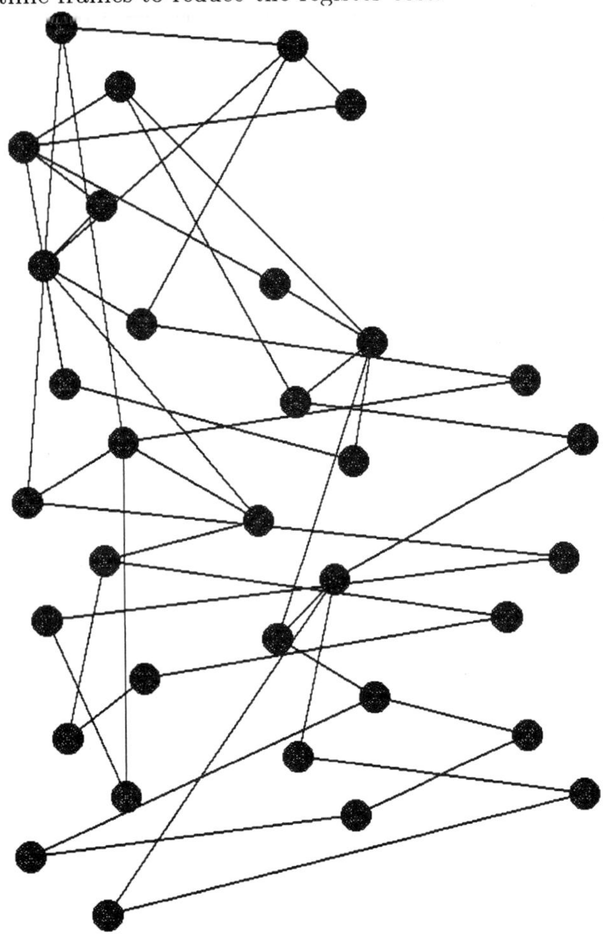

Figure 1: 5th *order filter* data flow graph *DFG*

Therefore, we propose an **original and general approach**, based on the ILP formulation, which consists in partitioning the set of operations, each partition violating as few constraints as possible (either data dependency or resource ones) and being balanced in terms of ILP variables. Considering a simple DFG would not be satisfactory, as a DFG only reflects data dependency constraints (see for instance the $5^{th} order filter$' DFG in Fig. 1). Thus, our partitioning is based on a *reduced constraint graph* extracted from the ILP formulation, whose vertices represent operations and edges represent constraints between operations. Performing minimum edge cut partitioning creates partitions with low constraint dependencies. As each partition leads to an optimal partial schedule, the final schedule, obtained by the concatenation of the partial schedules, is a very good approximation of the optimal one.

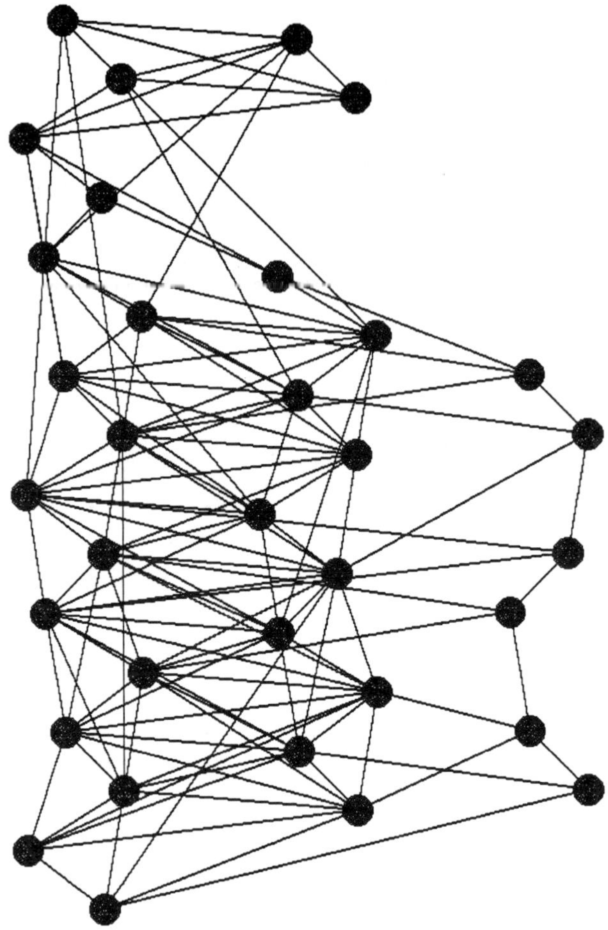

Figure 2: 5th *order filter* reduced constraint graph *RCG*

A $k-$partition

In order to determine the best value of k, one could iteratively try several decreasing values until it leads to an infeasible ILP formulation: starting with an $n-$partition, if all the partitioned formulations succeed to find a solution, try with a $(n-1)-$partition, and so on. This solution is realistic, as the computation times are largely decreased using the partitioning method (see below, and particularly the comparison between benchmarks which had a solution with the whole formulation and their partitioned solution). Moreover, one usually knows the approximate number of variables (N_{max}) that his solver can handle. Thus, an efficient solution is to determine directly, as a starting value, a number of partition which has a good chance to be the optimal (for example it would be $\left\lceil \frac{\sum_i L_i - S_i + 1}{N_{max}} \right\rceil$ with the scheduling problem).

In our specific scheduling problem [15], we have to

benchmarks	Nb equations	Nb variables	Nb operations	NB cycles	CPU time
2 way method partition 1	30	36	4	3	1.1s
2 way method partition 2	24	24	3	2	0.2s
Fast Fourier Transform partition 1	32	50	4	1	0.4s
Fast Fourier Transform partition 2	45	54	6	3	0.5s
Differential equation partition 1	63	77	10	3	0.7s
Differential equation partition 2	56	76	8	3	1.4s
3-way method partition 1	128	187	15	6	6 hr 34 min
3-way method partition 2	102	118	11	5	12 min 56 s
3-way meth. Classic partition 1	57	68	16	6	5.7s
3-way meth. Classic partition 1	33	33	9	4	0.9s
$5^{th}O$ Elliptic wave filter partition 1	121	204	15	7	1 hr 46 min
$5^{th}O$ Elliptic wave filter partition 2	135	195	17	4	2 hrs 58 min
$5^{th}O$ Elliptic wave filter partition 3	128	175	18	5	2hrs 20 min
$5^{th}O$ EWF Classic partition 1	65	65	17	10	7.9s
$5^{th}O$ EWF Classic partition 2	51	68	17	7	2.0 s

Table 2: ILP results using LP_SOLVE after ILP based partitioning.

deal with the number systems used for representing operands: the ILP formulation aims at finding the best (redundant or non redundant) encoding. As the outputs of a partition become the inputs of the following partition, and as the encoding of the primary inputs influence the scheduling, this implies that all the precedent partitions must have already been scheduled. This communication problem is also crucial when dealing with resource allocation and scheduling: the way operators share registers in a partition highly influences the resource sharing in the following partitions. Thus, these informations have to be brought to the other partitions. Therefore, it seems more efficient to make successive bi-partition after each local schedule rather than an initial global $k-$partition.

Considering the data flow graph $DFG = (V, E)$ such that the ILP formulation related to DFG could not be solved, The $k-$partition of $DFG = (V, E)$, with k as small as possible, defines k data flow graphs $DFG_i = (V_i, E_i)$, such that $V = V_1 \cup V_2 \cup \ldots \cup V_k$ and $V_i \cap V_j = \emptyset$ if $i \neq j$.

We are dealing with partition i
DFG_1 to DFG_{i-1} have been scheduled.
A reduced constraint graph, RCG_i is built:
  $RCG_i = (RCV_i, RCE_i, W_i)$, where
    $RCV_i = V \setminus \{V_1 \cup V_2 \cup \ldots \cup V_{i-1}\}$.
    RCE_i andW_i are defined in the next section.
    W_i aims at representing the "ILP size".
RCG_i is bipartitioned according to the ILP sizes:
  $S(RCG_i)$ refers to the "ILP size" of graph RCG_i.
  Thus, $S(RCG_i) = \sum_{v_j \in RCV_i} w(v_j)$,
    where $w(v_j)$ is the weight of v_j in RCG_i
  We create a bi-partition of RCG_i **with minimum edge cut**, of respective size,
    $\frac{S(RCG_i)}{k-i}$ and $\frac{(k-i-1).S(RCG_i)}{k-i}$.

Partitioning with minimum cut is known to be a NP-complete problem [7], but there are some efficient heuristics [6, 12]. Our method has been implemented using the Fiduccia and Mattheyses heuristic, which is an improvement of Kernighan and Lin Min-Cut heuristic.

As edges represent constraints, the idea behind min-cut partitioning is to minimize the constraint violations. Thus, this algorithm is efficient if RCG is a good representation of the different constraints. We will now address the problem of the reduced constraint graph definition.

3.2 Reduced constraint graph

The previous method is not specific to any particular problem. It has been tested with our scheduling problem, but it could also be used to any other ILP based problem. Therefore, the definition of the reduced constraint graph, RCG, should not depend on any particular problem. Nevertheless, there are a few observations that the graph should match.

- The input of the ILP formulation is a DFG. Thus, the graph vertices represent operations of the DFG.
- Our goal is to create partitions whose ILP formulations would take comparable computation times. As every operation does not have the same influence over the ILP computation time, the vertices must have a weight $w(e_j)$ reflecting their influence over this computation time.
- Edges must represent **any** constraint. In fact, this solution should not even be related to a scheduling problem, but, more generally, to the problem of resolving large ILP formulations.

We based our solution on a graph used by Pan, Dong and Liu [18] to solve a problem of constraint reduction in symbolic layout compaction: from a set, S, of linear programming constraints of the form $x_i - x_j \geq b$ (we will say that $x \in cn$ if variable x appears

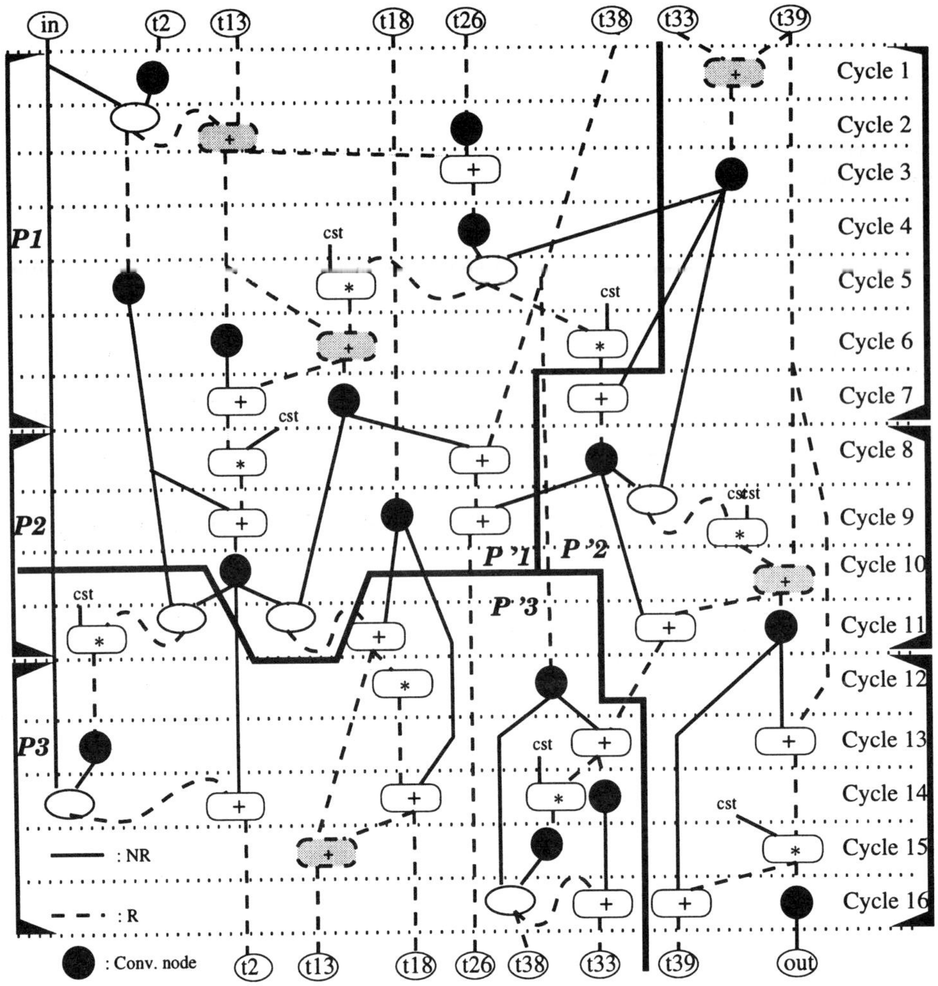

Figure 3: Result of the 5th *order filter* scheduling after a 3-partition (partitions are P1, P2 and P3 with a constraint graph partition, and P'1, P'2 and P'3 with a DFG partition).

in constraint cn), they create a directed graph, $G = (V, E)$, such that each variable x_i which appears in S is related to a vertex $v_i \in V$, and such that each constraint $cn : x_i - x_j \geq b, cn \in S$ is related to an edge $e : v_i \mapsto v_j, e \in E$, of weight b. From this graph, they solve a problem of subgraph reduction (i.e finding an equivalent graph with less edges).

We have extended this representation to ILP: from a constraint $\sum_{i,j} a_{i,j}.x_{i,j} \geq b$, where $x_{i,j}$ represents a variable related to operation o_i, we construct a complete graph $CG = (CV, CE)$, where a vertex $v_{i,j} \in V$ is related to the ILP variables $x_{i,j}$. It makes a constraint graph **of variables**. From this graph, we perform a coarsening phase which creates clusters $C_i = \{x_{p,q} | p = i\}$ (C_i contains all the variables related to operation o_i), in order to get a graph **of operations**. This defines a *reduced constraint graph* $RCG = (RCV, RCE, W)$ as follows:

From a set, ILP, of ILP constraints,

- $\forall i$, if $\exists cn \in ILP, j \in N | x_{i,j} \in cn$, we construct a vertex $v_i \in RCV$, weighted by $w(v_i) = Card_j(x_{i,j})$.
- If $\exists cn \in ILP, i_1, i_2, j_1, j_2 \in N | i_1 \neq i_2, x_{i_1,j_1} \in cn, x_{i_2,j_2} \in cn$, we construct an edge $e_{1,2} \in RCE$ between v_{i_1} and v_{i_2}.

This definition fits the previous observations, as RCG is an operation graph, whose vertices are weighted by the number of ILP variables linked to an operation, which has a great influence over the ILP computation time. Furthermore, the edges are constructed by each ILP constraint, explicitly and equally treated. This method could even be used for other problems than scheduling ones. Fig. 2 shows the reduced constraint graph for the 5th *order filter* design.

4 Results

We have tried this solution with the $5th$ *order elliptic wave filter*, using a 3-partition, for our scheduling and operator type selection problem. The 5*th order filter* 3-partition of Fig 3 has been obtained. It defines partitions P1, P2 and P3 and the resulting schedule is optimal for our specific problem. Compared to our partitions, the partitions P'1, P'2 and P'3 were obtained with a 3-partition based upon the DFG. The DFG based partition could not be exploited, as the partition is not "horizontal". This shows that, even if our method does not introduce any information specific to a scheduling method, it gives exploitable solutions.

Examples that did not need partitioning have also been tested, in order to get an idea of the degradation compared to the optimal. Only two examples had more cycles than the optimal: the differential equation design and the $5^{th} order filter$ EWF one, in its classical version, without redundant operators and operator type selection (3+3 instead of 5 and 10+7 instead of 16). However, in each case, the extra cycle is due to the junction between the two partitions, and a method closed to one presented in section 2 manages to find the optimal schedule. This solution consists in making a formulation of the whole design where the frames of the second partition operations are reduced to $\{T(i)-1, T(i)\}$ (where $T(i)$ is the schedule obtained by the partitioning solution), whereas first partition operations are fixed to the partitioning solution schedule. This solution allows to use the available resources of the last cycle of partition 1, and can be considered as a "smart" concatenation. All the others examples managed to find the optimal schedule. Besides, on every benchmark, the CPU time is largely decreased (see table 2 compared to table 1). This is particularly impressing with large examples (10 and 28 times faster for the 3-way method and the $5^{th} order filter$).

We have experimented this method with other examples, and a problem occurred when the operation time frames are too large. In such a case, the resource constraints become too numerous compared to data dependency ones, and the partitions obtained can not be exploited (for example, additions and multiplications are separated without regards of the precedence). This problem disappears if we manage to reduce the initial operation frames. A better solution would be to have a balanced influence of the different kinds of constraints. Therefore we plan to introduce a weight on the constraint graph edges. We are currently experimenting weighting functions which would balanced the influence of resource and data dependency constraints, even with large operation frames. Of course, this solution is not general anymore, and becomes specific to a scheduling problem.

5 Conclusion

To overcome the problem of ILP complexity, we have defined a partitioning methodology which constructs a reduced constraint graph $RCG = (V, E, W)$ to successively bi partition an ILP formulation. V represents the DFG nodes. E is built according to all the ILP constraints. W represents the ILP size.Therefore, this graph can be used to partition any ILP formulation based on a DFG, which are common in high level and system level synthesis. This method have been tested for the scheduling problem. It seems that a better partition can be obtained by weighing the edges of RCG according to the type of constraints (resource sharing or data dependency). We are currently working on this weight definition.

References

[1] A. Bender. Milp based task mapping for heterogeneous multiprocessor systems. In *proceedings of the European Design Automation Conference 96 (EuroDAC'96)*, 1996.

[2] N. N. Binh, M. Imai, and A. Shiomi. A new hw/sw partitioning algorithm for synthesizing the highest performance pipelined asips with multiple identical fus. In *proceedings of the European Design Automation Conference 96 (EuroDAC'96)*, 1996.

[3] S. Chaudhuri, S. A. Blythe, and R. A. Walker. An exact methodology for scheduling in a 3d design space. In *Proceedings of the International Symposium on System synthesis (ISSS 95)*, 1995.

[4] F. Depuydt, G. Goossens, and H. De Man. Clustering techniques for register optimization during scheduling preprocessing. In Louise Goto, Satoshi; Trevillyan, editor, *Proceedings of the IEEE International Conference on Computer-Aided Design*, pages 280–283, Santa Clara, CA, November 1991. IEEE Computer Society Press.

[5] Y.G DeCastelo-Vide e Souza, M.Potkonjak, and A. C. Parker. Optimal ilp-based approach for throughput optimization using simultaneous algorithm/architecture matching and retiming. In *Proceedings of the 32nd Design Automation Conference*, 1995.

[6] C.M. fiduccia and R.M. Mattheyses. A linear-time heuristic for improving network partitions. In *Proceedings of the 19th Design Automation Conference*, 1982.

[7] M.R. Garey and D.S. Johnson. *Computers and Intractability: A Guide to the Theory of NP-Completeness*. Ed. W.H. Freeman and Co., New-York, 1979.

[8] C. H. Gebotys and M. I. Elmasry. Optimal synthesis of high-performance architectures. *IEEE Journal of Solid-State Circuits*, 27(3), 92.

[9] C-T. Hwang and Y-C. Hsu. Zone scheduling. *IEEE Transaction on Computer Aided Design of Integrated Circuits and Systems*, 12(7):926–934, 1993.

[10] C-T. Hwang, J-H. Lee, and Y-C. Hsu. A formal approach to the scheduling problem in high level synthesis. *IEEE Transaction on Computer Aided Design*, 10(4):464–475, April 1991.

[11] I. Karkowski and R.H.J.M Otten. An automatic hardware/software partitioner based on the possibilistic programming. In *proceedings of the European Design and Test Conference 96 (EDTC'96)*, 1996.

[12] B.W. Kernighan and S. Lin. An efficient heuristic procedure for partitioning graphs. *Bell System Technical Journal*, 49, 1970.

[13] B. Landwehr, P. Marwedel, and R. Dömer. Oscar: Optimum simultaneous scheduling allocation and resource binding based on integer programming. In *proceedings of the EuroDAC'94*, 1994.

[14] R. Leupers and P. Marwedel. Time-constrained code compaction for dsps. In *proceedings of International Symposium on system synthesis 95 (ISSS'95)*, 1995.

[15] A. Mignotte and O. Peyran. Scheduling using mixed arithmetic: an ilp formulation. In *proceedings of the European Design and Test Conference 1997*, 1997.

[16] Anne Mignotte, Jean-Michel Muller, and Olivier Peyran. Mixed arithmetics: Introduction and design structure. In *proceedings of MPCS'96*, 1996.

[17] R. Niemann and P. Marwedel. Harware/software partitioning using integer programming. In *proceedings of the European Design and Test Conference 96 (EDTC'96)*, 1996.

[18] Peichen Pan, Sai-Keung Dong, and C.L. Liu. Optimal graph constraint reduction for symbolic layout compaction. In *Proceedings of the 30th Design Automation Conference*, 1993.

[19] S. Prakash and A. C. Parker. Sos: Synthesis of application-specific heterogeneous multiprocessor systems. *Journal of Parallel and Distributed Computing*, 16:338 – 351, 1992.

[20] M. Schwiegershausen and P. Pirsch. A system level design methodology for the optimization of heterogeneous multiprocessors. In *proceedings of the International Symposium on System Level Synthesis 95 (ISSS'95)*, 1995.

[21] H. Tomiyama and H. Yasuura. Otpimal code placement of embedded software for instruction caches. In *Proceedings of the European Design and Test Conference (EDTC'96)*, 1996.

A New Optimization Technique for Improving Resource Exploitation and Critical Path Minimization

Birger Landwehr, Peter Marwedel

Dept. of Computer Science XII, University of Dortmund
D-44221 Dortmund, Germany

Abstract

This paper presents a novel approach to algebraic optimization of data-flow graphs in the domain of computationally intensive applications. The presented approach is based upon the paradigm of simulated evolution which has been proven to be a powerful method for solving large non-linear optimization problems. We introduce a genetic algorithm with a new chromosomal representation of data-flow graphs that serves as a basis for preserving the correctness of algebraic transformations and allows an efficient implementation of the genetic operators. Furthermore, we introduce a new class of hardware-related transformation rules which for the first time allow to take existing component libraries into account. The efficiency of our method is demonstrated by encouraging experimental results for several standard benchmarks.

1 Introduction

The very first step in the design-flow of digital systems is concerned with formulating the behavioral specification in a hardware description language such as VHDL [9]. This behavioral description forms the basis for all subsequent design steps starting with behavioral (or high-level) synthesis at the algorithmic level. High-level synthesis deals with the transformation of the behavioral description into a netlist of RT-level components and is generally understood as a mapping of operations of the data-flow graph (DFG) to control steps and to suitable components of an existing library [16]. Since high-level synthesis systems directly operate on the internal representation of the behavioral description it is quite obvious that the chosen formulation style has a lasting effect to later design steps and therefore to the final result.

Although the use of high-level synthesis systems has gained acceptance during the recent years, the actual question of how to suitably formulate a behavioral description has often been underrated. This question particularly arises in the domain of digital signal applications which are characterized by complex arithmetical computations resulting in complex data-flow graphs.

Figure 1 demonstrates the effect of different transformations for expression $((a * c) + (b * c)) + d$ with respect to time and resource equirements.

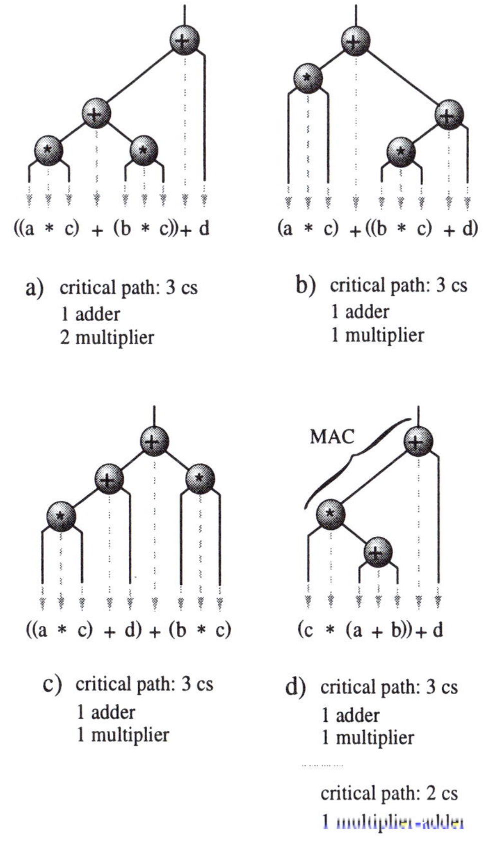

Figure 1: Time and resource requirements for equivalent expressions

1080-1820/97 $10.00 © 1997 IEEE

Figure 1a presents the original expression requiring one adder and two multipliers with a critical path of 3 control steps[†]. After applying the associativity law (figure 1b), the order of operations has been changed such that component sharing could be improved. Therefore, only one multiplier is required after the transformation. The additional exploitation of commutativity shown in figure 1c neither leads to a further improvement of resource sharing nor to a shorter critical path, but may increase the applicability of other transformation rules. Figure 1d shows the expression after exploiting distributivity. Although implementation costs and critical path have not further changed, the application of a special component-driven transformation $(x * y) + z \Leftrightarrow \text{MAC}(x, y, z)$ is possible. This rule implies a mapping of the two operations to a single multiplier-adder-accumulator (MAC) in contrast to a separate implementation by one multiplier and one adder. As we can recognize, the use of this component-directed transformation has a major impact on the synthesis result: in the latter case, only one MAC is required because this component can also be employed as a simple multiplier or adder by applying the particular identity element to the corresponding input port. Since components like MACs are capable of performing two ore more operations in one execution step we call such functional units *complex components* or *BIC* (*built-in-chaining*)*-components* [15].

Even this simple example has shown that synthesis results are strongly dependent on the choice of a certain formulation style of the behavioral description. Obviously, it might be difficult to recognize in advance how to formulate a behavioral description that leads to the best synthesis result. This becomes almost impossible for complex data-flow dominated circuits. The goal of applying algebraic transformations can thus be stated as *making synthesis results mostly independent of the formulation style* or *to transform the input description such that synthesis yields better results than for the original description.*

The remainder of this paper is organized as follows: Section 2 gives an overview of the research on algebraic optimization in different areas. After introducing some hardware-related transformations in section 3 we describe the genetic algorithm including the chromosomal representation and the genetic operators in section 4. Section 5 presents experimental results for several standard benchmarks, and section 6 concludes the paper.

2 Related work

The use of algebraic transformations has been established in different domains: In classical computer-algebra systems such as MAPLE [5] or MATHEMATICA [21] they are indispensable for the transformation and simplification of algebraic expressions. In the domain of high-level-language compilers (see [2] for an overview) algebraic transformations are mainly used for tree height reduction, common subexpression elimination, constant folding, constant propagation and rather simple optimizations based on strength reduction.

In the area of high-level synthesis algebraic transformations have been particularly used for improving resource utilization [18] [17], tree-height minimization [6] [7], the maximization of data throughput [8] [10] and minimization of power consumption [3]. The use of algebraic transformations in combination with complex components (e.g. MACs) was first proposed in [14]. However, to the author's knowledge there is no optimization technique that exploits component-directed transformations in the same extent as the approach presented in this paper.

An approach based upon evolutionary programming for an area efficient design of Application Specific Programmable Processors (ASPP) has been published in [22]. ASPPs are programmable architectures which are designed for a set of different algorithms. The underlying approach is based on a genetic algorithm for transforming the particular data-flow graphs such that a given *behavioral kernel* (defined by a set of RT-level components) is optimally exploited by the algorithms.

Concerning the chromosomal representation of data-flow graphs and genetic operators (mutation and crossover), our method appears similar to [22]. However, it combines the concept of evolutionary programming with the algebraic optimization techniques for critical path minimization and improved resource exploitation on a finer level of granularity.

3 Overview of algebraic optimizations

The introducing example has already shown that apart from the exploitation of algebraic laws e.g. commutativity, associativity, distributivity etc., even hardware-related transformations have the potential of a considerable reduction of hardware costs and the critical path. In this section we demonstrate how hardware-related transformations can be specifically employed in order to reduce resource requirements.

3.1 General hardware-related transformations

First let us consider the expression $x * c$. Let c be a constant with $c = 2^n$. Obviously, the expression can be implemented in several ways: a) using a multiplier, b)

[†] In this example all operations are assumed to be single-cycled.

using a shifter, or c) by an offset in the wiring pattern. The transformation of the multiplication to a possibly hardwired shift-operation is well known as *strength reduction* [2], i.e. an expression with a cost intensive realization is always replaced by a "cheaper" expression. This optimization technique is usually exploited in almost all modern software compilers, however rarely in synthesis systems.

The exploitation of powers of two is not restricted to multiplications: Consider the expression $x_{[\text{msb}:0]} + y$. [msb:0] represents the bit-slice[†] of x, and $y = 2^n$ is a constant again. Since an addition with 2^n increments only the upper bits [msb:n] whereas the lower bits [n-1:0] are not affected, we can state the following transformations: $x + 2^n \Leftrightarrow (x_{[\text{msb}:n]} + 1) \,\&\, x_{[n-1:0]} \Leftrightarrow \text{inc}(x_{[\text{msb}:n]}) \,\&\, x_{[n-1:0]}$. Hence, additions with a power of two can be replaced by increment operations and thus lead to cheaper hardware realization. The same transformations also hold for $x - 2^n$, however applying a decrement operation. The reduction of an addition to an increment operation can also be exploited in order to find a cost efficient implementation for $1 - z$ with only one incrementer and one inverter: $1-z \Leftrightarrow -z+1 \Leftrightarrow \overline{z}+2 \Leftrightarrow \text{inc}(\overline{z}_{[\text{msb}:1]}) \,\&\, z_{[0:0]}$.

All transformations presented above are accompanied by an immediate reduction of the realization costs. However, we still have to take transformations into account which temporarily have the opposite effect:

Constant Unfolding is a technique that promises a further improvement in terms of cost and speed by splitting constants into a power of two and a remainder. Consider the rule $c = [c - r]^{\ddagger} + r$, where $[c - r] = 2^n$ is a constant. We can split expression $9 * x$ into $(2^3 + 1) * x$, which is equivalent to $2^3 * x + x$. Since $2^3 * x = x \,\&000$ can be implemented by an offset in the wiring pattern, only one adder instead of a multiplier is required to realize the expression. Another transformation rule is concerned with introducing identity elements which may be necessary to increase the applicability of further transformations: $a + a \Leftrightarrow (a * 1) + (a * 1) \Leftrightarrow a * (1 + 1) \Leftrightarrow a * 2$. Obviously, the use of identity elements is necessary for the formal proof of equivalence without knowledge of the rule $a + a \Leftrightarrow a * 2$. Although the introduction of identity elements helps to find new formulations of an initial expression, it also temporarily increases the implementation costs. Concerning the length of the critical path it was shown in [13] that the creation of additional operations in the DFG may have a positive effect on synthesis results. Therefore it is essential that the underlying optimization method does not reject transformations which temporarily lead to suboptimal results.

[†]Corresponding to the VHDL [9] notation `x : Bit_vector (msb DOWNTO 0)` where `msb` denotes the most significant bit, i.e. `msb := x'LENGTH - 1`.

[‡]The brackets mean an instant evaluation of the subexpression, e.g. constant $c = 9$ is splitted into $[9 - 1] + 1 = 8 + 1$.

3.2 Component-driven transformations

As we saw in the introduction, the use of a special component-driven transformation for employing multiplier-adders has the potential of a further reduction of both hardware costs and the critical path. In the following example we demonstrate how such transformations can be applied in order to exploit existing library components more efficiently: Consider the transformation rule $x + y + z \Leftrightarrow \text{CADD}(x, y, z)$, with $z \in \{0, 1\}$. This rule implies that expression $x + y + z$ is mapped to one carry-adder. In combination with the transformations presented above the following rule allows to implement $x + y - z$ with $y = 2^n$ by only one carry-adder and one inverter: $(x + \overline{z}_{[\text{msb}:n]} + 1) \,\&\, \overline{z}_{[n-1:0]} \Leftrightarrow \text{CADD}(x, \overline{z}_{[\text{msb}:n]}, 1) \,\&\, \overline{z}_{[n-1:0]}$.

This section has shown the applicability and the positive impact of hardware-related transformation rules concerning hardware costs and speed. We also have recognized that some transformations temporarily may increase the costs but allow the application of further rules which globally may decrease the costs. Since cost-driven heuristics do not work appropriately in this case, we formulated the problem using a probabilistic approach based on a genetic algorithm which will be presented in the following section.

4 Algebraic optimization by genetic algorithm

The general principle in natural evolution as well as evolutionary algorithms is the optimization of a population's fitness in the course of generations driven by the randomized processes of *selection*, *recombination* and *mutation*. Genetic algorithms as one representative of evolutionary algorithms (see [1] for an overview) have been proven to be very powerful for searching vast solution spaces. Solutions found by genetic algorithms are generally close to the global optimum.

4.1 Chromosomal representation

Each chromosome of the population represents one semantically equivalent formulation of an initial data-flow graph. The genes which are located on the chromosome—or the gene positions, to be precise—represent the operations of the DFG together with references to the predecessor operations.

Example: We use expression $((a * 2) + (b * 2)) + 1$ as a running example in order to demonstrate the chromosomal representation and the particular genetic operators. Figure 2 depicts the chromosome representation $\alpha \rightarrow \beta \rightarrow \gamma \rightarrow \delta$ with its genes $\alpha : \beta + 1, \;\; \beta : \gamma + \delta, \;\; \gamma : a * 2,$ and $\delta : b * 2.$

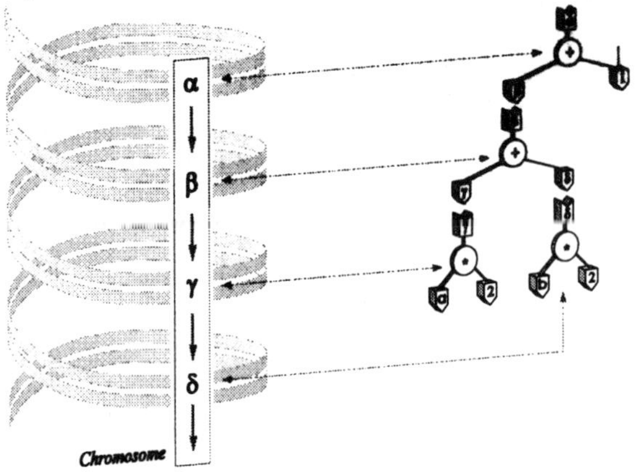

Figure 2: Representation of expression $((a * 2) + (b * 2)) + 1$ by a sequence of genes

Each gene has a specific phenotype, called its allele. We distinguish different alleles of the same gene by roman numbers I, II, etc. Operations of the original data-flow graph are thus represented by the alleles $\alpha_{\text{I}}, \beta_{\text{I}}$ etc. The set of alleles for one specific gene represents the set of functionally equivalent expressions. Since all alleles of the same gene are semantically equivalent they are mutually replacable without changing the functionality of the entire DFG. Nevertheless, the resulting DFG can be distinguished by potentially different hardware costs and critical path lengths.

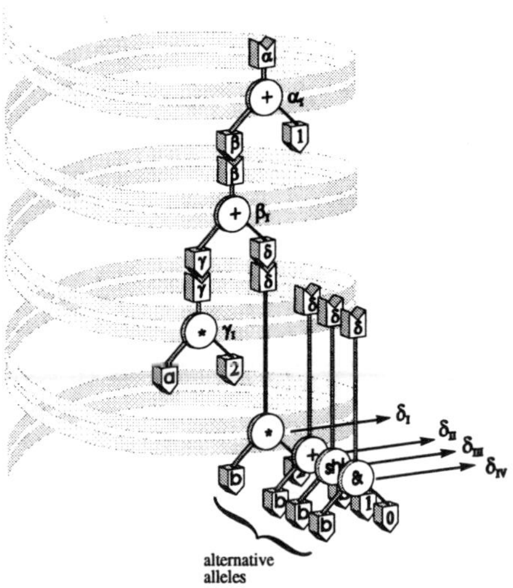

Figure 3: Alternative alleles

Figure 3 shows new alternative alleles for the gene δ after applying the transformations $b * 2 \Rightarrow b + b \Rightarrow b$ shl $1 \Rightarrow b$ & 0. Obviously, each allele of gene δ can be replaced by any other allele without changing the semantics of the expression. In the same way, algebraic transformations can be applied to the other genes of the chromosome.

Figure 4 presents the gene pool for our example after applying associativity, distributivity and the demonstrated simplification of multiplications. The gene pool serves as a basis for the creation of the initial population: for each chromosome that represents one individual of the population we can arbitrarily select one allele at each gene position (figure 4). Due to the fact that each chromosome implicitly represents the structure of a data-flow graph, the creation of any DFG can be performed in linear time $\text{O}(n)$. In this example, the chromosome $\alpha_{\text{III}} \rightarrow \beta_{\text{III}} \rightarrow \gamma_{\text{IV}} \rightarrow \delta_{\text{II}} \rightarrow \epsilon_{\text{II}} \rightarrow \zeta_{\text{I}}$ represents the expression inc$((a + b)$ shl $1)$. It should be mentioned that chromosomes may also contain some *redundant* genes (in this example: $\gamma \rightarrow \delta \rightarrow \epsilon$) which have no direct influence on the created DFG. However, redundant genes can be reactivated instantly by small mutations of the chromosome.

As we have seen, the chromosomal representation introduced above guarantees that all subexpressions referenced by different alleles at the same gene position are semantically equivalent. This means that every possible allele substitution at any gene position will subsequently lead to a new semantically equivalent data-flow graph. This property is crucial for preserving the correctness of the used genetic operators, namely *mutation* and *crossover*.

4.2 Genetic operators

4.2.1 Mutation

The principle of mutation was implicitly shown in figure 3: Without changing the entire semantics we can transform a data-flow graph by a simple gene mutation that substitutes the selected allele by another one at the same gene position. For example, the substitution of allele δ_{I} by δ_{IV} represents the transformation $(a*2)+(b*2)+1 \Leftrightarrow (a*2)+(b \;\&\; 0)+1$.

Obviously, the mutation operator can be implemented in time O(1).

4.2.2 Crossover

The goal of crossover is to recombine the parental properties and its transmission to the new offspring. In the meaning of transforming algebraic expressions, crossover recombines subexpressions of the parental data-flow graphs and can be sketched as follows:

1. Create two new chromosomes representing the children.

2. Select an arbitrary gene position.

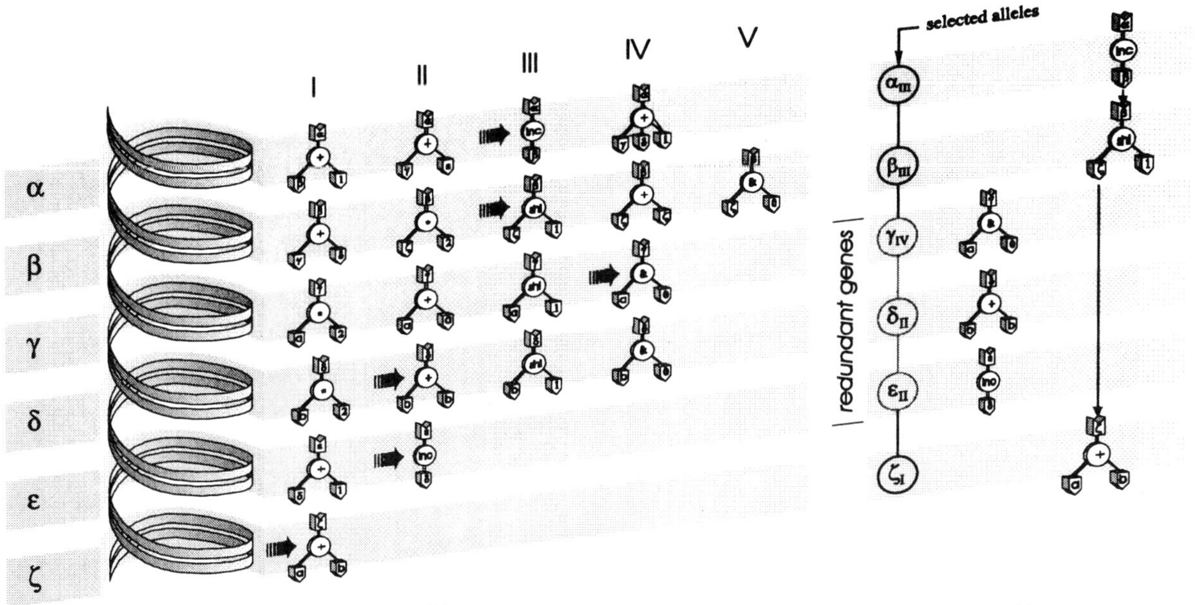

(a) (b)

Figure 4: (a) Extended gene pool for the running example and
(b) exemplary creation of a DFG by selecting alleles at each gene position

3. Copy all alleles from first (second) chromosome up to the selected gene position to the first (second) child.

4. Copy the remainder of the first (second) chromosome the second (first) child.

Also crossover benefits from the underlying representation and always creates only those DFGs which are semantically equivalent to the initial specification. Obviously, crossover can be implemented in linear time $O(n)$ where n represents the chromosome length.

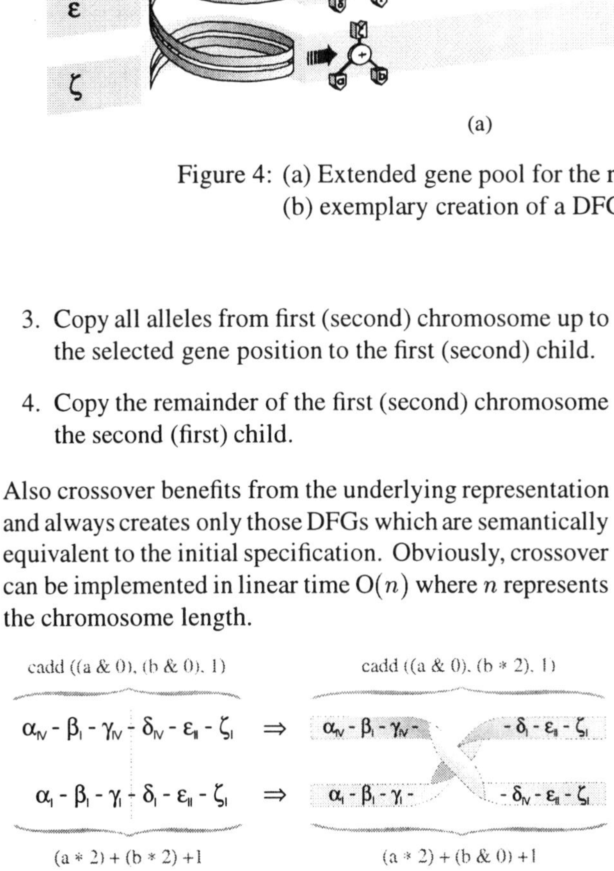

Figure 4: Crossover

Figure 4 demonstrates crossover for our running example. As initial chromosomes we use $\alpha_{IV} \longrightarrow \beta_I \longrightarrow \gamma_{IV} \longrightarrow \delta_{IV} \longrightarrow \epsilon_{II} \longrightarrow \zeta_I$ corresponding to cadd$((a \; \& \; 0), (b \; \& \; 0), 1)$ and $\alpha_I \longrightarrow \beta_I \longrightarrow \gamma_I \longrightarrow \delta_I \longrightarrow \epsilon_{II} \longrightarrow \zeta_I$ which corresponds to $((a * 2) + (b * 2)) + 1$. The crossover position has been chosen at gene position δ. In the resulting expressions, $(b \; \& \; 0)$ and $(b * 2)$ have been interchanged.

4.2.3 Selection

Selection is a crucial process in (simulated) evolution that favors individuals of higher fitness to survive ("survival of the fittest") and thus become the co-founders of the next population. Generally, we presume the probability of an individual to be selected is proportional to its fitness. This enables even individuals with a lower fitness to be selected and thus to transmit their gene information to the offspring.

In the meaning of the final hardware realization we define the fitness as weighted sum of the required functional units and the length of the critical path. In contrast to the critical path computation that can be done in linear time by an ASAP (or ALAP) scheduling, the exact computation of resource costs is NP-complete. Therefore, we have to employ resource estimation techniques (e.g. [19]) in order to value the effect of performed transformations. Surprisingly, experimental results have shown that even simple fitness functions are sufficient for producing excellent optimization results (see figure 5–6 and table 2). We used a combination of the critical path length and hardware costs computed by *direct compilation*, i.e. each operation of the data-flow is associated with certain hardware costs. An advantage of using direct compilation is its efficient implementation in linear time and is thus crucial for the fast execution time of the optimization routine.

4.3 Skeleton of the genetic algorithm

Figure 5 presents the genetic algorithm that serves as a basis for algebraic optimization and takes pattern from the standard algorithm in [4].

```
1       initialize individuals of the population p
2       FOR EACH epoch e DO
2.1       apply transformation rules to the current population p
2.2       FOR EACH generation g DO
2.2.1       compute fitness of all individuals
2.2.2       select individuals according to their fitness
2.2.3       create offspring by crossover
2.2.4       mutate offspring
2.2.5       replace individuals of the current population by the offspring
2.2.6       exit loop, if criterion T_G is fulfilled
2.2'      END
2.3       terminate, if criterion T_E is fulfilled
2'      END
```

Figure 5: Skeleton of the genetic algorithm

The algorithm consists of an outer and an inner loop. The inner loop repeats the tasks of fitness computation, selection, crossover, mutation and replacement of individuals as long as the loop-exit criterion T_G is not fulfilled. The loop exit is usually controlled by the state of the generation counter or by the yielded gain of the population's fitness.

The outer loop is required in order to extend the current gene pool in two directions: On the one hand we introduced the new alleles δ_{II}, δ_{III} and δ_{IV} in our example to represent the transformations $a * 2 \Rightarrow a + a \Rightarrow a \text{ shl } 1 \Rightarrow a \ \& \ 0$. On the other hand the chromosome is extended by new genes along with their alleles. In our example, we introduced gene ϵ with its alleles ϵ_I and ϵ_{II} and the new allele α_{II} in order to exploit the associativity.

In contrast to the actual composition of new DFGs including their recombination, mutation and selection in progress with the *generations*, we call the continuous extension of the current gene pool by transformations *epochs*. The termination of the entire algorithm is controlled by fulfilling T_E that can be realized in the same way as the loop-exit criterion T_G.

5 Experimental results

We applied the presented algorithm to several standard benchmarks. For each example we achieved a gain of up to 30 % concerning the critical path and an area gain of up to 26 %. On the basis of empirical tests we determined the following genetic parameters: population size: 80 individuals; number of individuals in the population to be replaced by the offspring: 60; number of generations: 40, Mutation rate: 0.1.

All presented results have been computed on a SparcStation 20. The execution time of the optimization routine was for all examples approximately one minute for the chosen parameters.

Figure 6 and 7 present the initial and the optimized dataflow graph of the 5th-order elliptical wave filter† [11], respectively. All operations are assumed to be single-cycled. The *macro-nodes* in fig. 7 represent multiply-add subexpressions which have been bound to MACs. These complex components can also be used for computing additions by simply applying the corresponding identity elements to the appropriate inports (control steps 3, 8, 9, 11 in fig. 8).

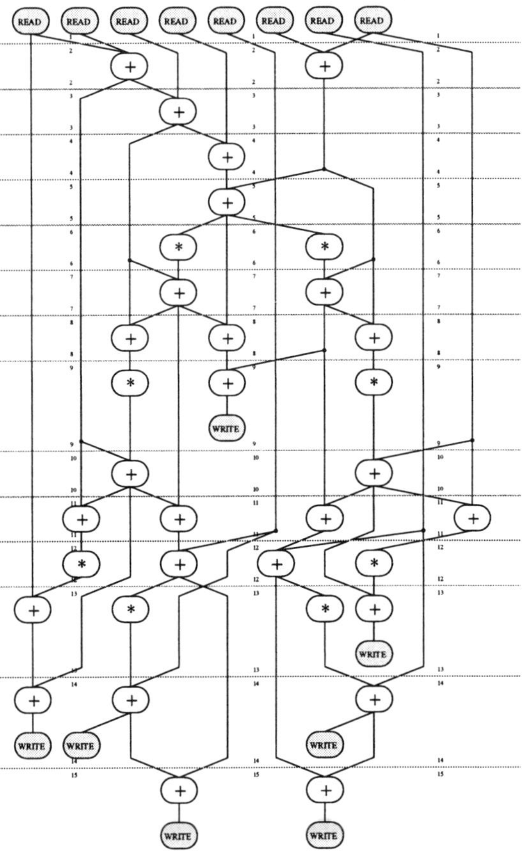

Figure 6: Initial DFG of the EWF-benchmark

†in contrast to some existing formulations of this benchmark containing only multiplications by 2, we dispense with the exploitation of the special transformation rule $x * 2 \rightarrow x \ \& \ 0$. I.e. we assumed multiplications with different coefficients in order to produce results for rather realistic applications

benchmark[a]	#cs	add $2405k\lambda^2$	sub $2433k\lambda^2$	mult $14717k\lambda^2$	mac $15435k\lambda^2$	gain (area[b]) $\frac{area_{orig}-area_{opt}}{area_{orig}}$	gain (#cs) $\frac{\#cs_{orig}-\#cs_{opt}}{\#cs_{orig}}$
EWF_{orig}	14	3	0	2	0		
EWF_{opt}	10	2	0	0	2	**2.6 %**	**28.6 %**
FFT_{orig}	8	3	4	12	0		
FFT_{opt}	8	0	4	5	4	**25.1 %**	0 %
FIR_{orig}	10	2	0	2	0		
FIR_{opt}	7	2	0	1	1	-2 %	**30.0 %**
$FDCT_{orig}$	6	4	2	8	0		
$FDCT_{opt}$	6	1	2	3	3	**26.1 %**	0 %
BF_{orig}	15	2	1	2	0		
BF_{opt}	14	0	1	0	2	**9.1 %**	**6.7 %**
$Edge_{orig}$	10	2	2	4	0		
$Edge_{opt}$	9	0	2	1	3	**3.9 %**	**10.0 %**

Table 2: Optimization results

[a]EWF: elliptical wave filter, FFT: fast fourier transformation, FIR: finite impulse response filter, FDCT: fast discrete cosine transformation, BF: bandpass filter, Edge: edge detection

[b]area of allocated functional units

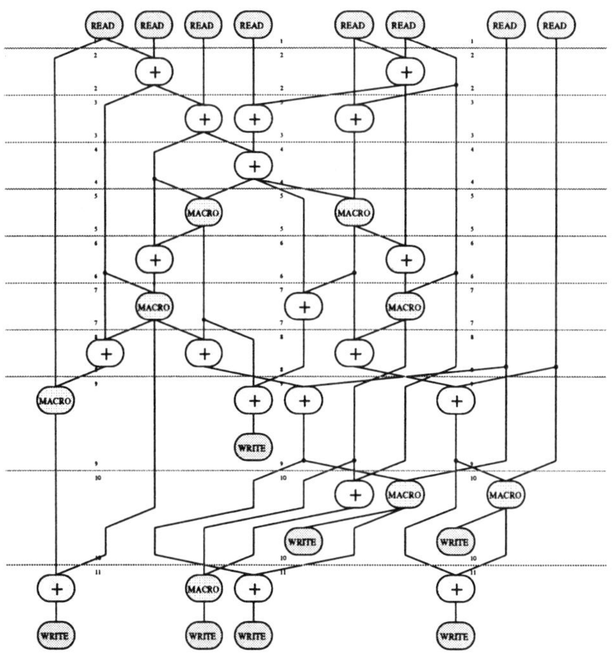

Figure 7: Optimized DFG of the EWF-benchmark

It seems to be quite obvious, that manually formulating a behavioral description with the same quality of the presented optimized DFG is virtually impossible. In this example, the critical path has been shortened from 14 to 10 control-steps requiring only two MACs and two adders instead of three adders and two multipliers.

Table 2 shows the synthesis results for several benchmarks before and after algebraic optimization. We used the high-level synthesis system OSCAR [12] for synthesizing the original and the optimized design. Due to its underlying integer programming formulation all presented results are optimal concerning the overall costs of functional units. Areas and delays[‡] of functional units have been adopted from the underlying 1.0μ VLSI component library [20]. The execution times of high-level synthesis were always less than one second.

6 Conclusion

We presented a genetic algorithm based approach for algebraic optimization of data-flow graphs. Due to the underlying chromosomal representation all genetic operators are correctness preserving and can be implemented very efficiently. Apart from standard transformation rules such as commutativity, associativity and distributivity, we also support hardware-related transformations. It has been shown that even these transformations have a positive effect to the quality of the achieved synthesis results. Since all rules are stored in an external library, they can be modified or extended by the designer.

The system has been implemented as a front end to an ILP-based synthesis system [12] and benefits from its capability of supporting complex component libraries. However, the approach can also be easily realized as a source-to-source (e.g. VHDL-to-VHDL) optimizer that

[‡]add [vdp1add001]: 16.1 ns , sub [vdp1sub001]: 16.7 ns, mult [vdp3mlt004]: 112.1 ns, mac [vdp3mlt006]: 112.1 ns

enables the employed synthesis system to support complex functional units (e.g. MACs).

The experimental results have shown the efficiency of our method. For all examined benchmarks (EWF, FFT, FDCT, FIR-Filter, Bandpass Filter, Edge Detection) we could achieve a considerable reduction of the critical path and/or the area.

References

[1] T. Bäck and F. Hoffmeister. Global Optimization by Means of Evolutionary Algorithms. *in: A. N. Antamoshkin (Ed): Random Search as a Method for Adaption and Optimization of Complex Systems, pages 17-21, Divnogorsk, UdSSR, March 1991, Krasnojarsk Space Technology University*, 1991.

[2] D. F. Bacon, S. L. Graham, and O. J. Sharp. Compiler Transformations for High-Performance Computing. *ACM Computing Surveys, Vol. 26, No. 4*, pages 345–420, 1994.

[3] A. P. Chandrakasan, M. Potkonjak, R. Mehra, J. Rabaey, and R. W. Brodersen. Optimizing Power Using Transformations. *IEEE Transactions on CAD, Vol. 14, No. 1*, pages 12–31, 1995.

[4] L. Davis. *Handbook of Genetic Algorithms*. Van Nostrand Reinhold, 1991.

[5] K. O: Geddes, G. H. Gonnet, and B. W. Char. MAPLE User's Manual (2nd ed.). Technical Report CS-82-40, University of Waterloo, 1982.

[6] R. Hartley and A. E. Casavant. Tree-Height Minimization in Pipelined Architectures. *Proceedings of the International Conference on Computer-Aided Design*, pages 112–115, 1989.

[7] R. Hartley and A. E. Casavant. Optimizing Pipelined Networks of Associative and Commutative Operators. *IEEE Transactions on CAD, Vol. 13, No. 11*, pages 1418–1425, 1994.

[8] S.-H. Huang and J. M. Rabaey. Maximizing the Throughput of High Performance Applications Using Behavioral Transformations. *Proceedings of the EDAC*, pages 25–30, 1994.

[9] Design Automation Standards Subcommittee of the IEEE. IEEE Standard VHDL Language Reference Manual (IEEE Std. 1076-87). *IEEE Inc., New York*, 1988.

[10] Z. Iqbal, M. Potkonjak, S. Dey, and A. Parker. Critical Path Optimization Using Retiming and Algebraic Speed-Up. *Proceedings of the 30th Design Automation Conference*, pages 573–577, 1993.

[11] S. Y. Kung, H. J. Whitehouse, and T. Khailath. *VLSI and Modern Signal Processing*. Prentice Hall, 1985.

[12] B. Landwehr, P. Marwedel, and R. Dömer. OSCAR: Optimum Simultaneous Scheduling, Allocation and Resource Binding Based on Integer Programming. *Proceedings of the EURO-DAC*, pages 90–95, 1994.

[13] D. A. Lobo and B. M. Pangrle. Redundant Operator Creation: A Scheduling Optimization Technique. *Proceedings of the 28th Design Automation Conference*, pages 775–778, 1991.

[14] P. Marwedel. Matching System and Component Behaviour in the MIMOLA Synthesis Tools. *Proceedings of the EDAC*, pages 146–156, 1990.

[15] P. Marwedel, B. Landwehr, and R. Dömer. Built-in Chaining: Introducing Complex Components into Architectural Synthesis. *Proceedings of the ASP-DAC*, 1997.

[16] M. C. McFarland, A. C. Parker, and R. Camposano. The High-Level Synthesis of Digital Systems. *Proceedings of the IEEE, Vol. 78, No. 2*, pages 301–318, 1990.

[17] M. Potkonjak and S. Dey. Optimizing Resource Utilization and Testability using Hot Potato Techniques. *Proceedings of the 31st Design Automation Conference*, pages 201–205, 1994.

[18] M. Potkonjak and J. Rabaey. Optimizing Resource Utilization by Transformations. *IEEE Transactions on CAD, Vol. 13, No. 3*, pages 277–292, 1994.

[19] A. Sharma and R. Jain. Estimating Architectural Resources and Performance for High-Level Synthesis Applications. *IEEE Transactions on VLSI Systems, Vol. 1, No. 2, June 1993*, 1993.

[20] VLSI Technology Inc. Library Manuals, 1993.

[21] S. Wolfram. *Mathematica, A System for Doing Mathematics by Computer*. Addison Wesley, 1988.

[22] W. Zhao and C. A. Papachristou. An Evolution Programming Approach on Multiple Behaviors for the Design of Application Specific Programmable Processors. *Proceedings of ED & TC*, 1996.

Session 4

Memory Management Issues

Fast and extensive system-level memory exploration for ATM applications

Peter Slock, Sven Wuytack, Francky Catthoor, Gjalt de Jong†

IMEC, Kapeldreef 75, B-3001 Leuven, Belgium
†Alcatel Telecom, Francis Wellesplein 1, B-2018 Antwerp, Belgium

Abstract

In this paper, our memory architecture exploration methodology and CAD techniques for network protocol applications are presented. Prototype tools have been implemented, and applied on part of an industrial ATM application to show how our novel approach can be used to easily and thoroughly explore the memory organization search space at the system-level. An extended, novel method for signal to memory assignment is proposed which takes into account memory access conflict constraints. The number of conflicts is first optimized by our flow-graph balancing technique. Significant power and area savings were obtained by performing the exploration thoroughly at each of the degrees of freedom in the global search space.

1 Introduction

The complexity of modern telecommunication systems is rapidly increasing. A wide variety of services has to be transported, and elaborate network management is needed. Such complex systems require a combination of hardware and software components in order to deliver the required functionalities at the desired performance level. For certain speed critical functionalities, such as packet handling and congestion control, it is necessary to implement the corresponding components in hardware using custom Application Specific Integrated Circuits. These ASICs are usually at the heart of modern telecommunication systems.

For applications in this domain, the desired behavior is often characterized by complex algorithms that operate on large dynamically allocated stored data structures (e.g. linked list, trees, dynamic FIFOs). This includes especially the transport layer in ATM networks and LAN/WAN technology. Ideally, the specification should reflect the "conceptual" partitioning of the problem, which typically correspond to abstract data types along with services provided on them, and algorithms for the different processing tasks. As these conceptual entities can be readily specified in an object-oriented programming model, using data abstraction and class inheritance features, we use the C++ programming language for the behavioral algorithmic specification as motivated in our global MATISSE approach[5].

For this type of applications, typically a (very) large part of the area cost is due to memory units. Also the power for data-dominated applications is heavily dominated by the storage and transfers [8]. Hence, we believe that a dominating factor in the system level design is provided by the organization of the global communication and data storage. Therefore we have proposed a design methodology in which the storage related issues (virtual and physical memory management) are optimized as a first step [5], *before* doing the detailed scheduling, and data-path and controller synthesis.

Two main tasks in our memory management methodology are flow graph balancing (FGB) and memory allocation/assignment. FGB is executed prior to the more conventional memory allocation/assignment tasks [1, 7, 12]. The goal of the task is to minimize the required memory bandwidth within the given cycle budget, by adding ordering constraints to the flow graph. This allows the subsequent memory allocation/assignment tasks to come up with a memory architecture with a small number of memories and memory ports. The CAD techniques to support these tasks will be detailed in this paper and their feasibility will be demonstrated by prototype tools.

The extensive design exploration which is feasible by applying our methodology and its heavy impact on storage area and power will be illustrated based on a representative module of an industrial ATM application, provided by Alcatel.

The rest of the paper is organized as follows. Section 2 presents the related work. In Section 3 our physical memory management methodology and the CAD techniques for two key tasks are presented. Section 5 illustrates the extensive exploration which can be obtained with the proposed methodology and the use of our prototype tools on a representative ATM example, introduced in Section 4. Section 6 contains the conclusions.

2 Related Work

Although behavioral hardware synthesis has been an active field of research for over one decade (see e.g. [2]), and commercial behavioral synthesis tools are currently emerging (e.g. Synopsys' Behavioral Compiler, Mentor Graphics' DSPStation, and Synthesia's SYNT system), support for complex data structures and memory synthesis related problems is generally limited. An exception to this is the video and image processing area for which several groups have proposed memory management support [7, 12] including also our own work on the ATOMIUM methodology and partial prototype tool support [9, 17]. This is however restricted to the support of statically declared arrays and records.

The data structures in the communication network protocol domain are however different, as indicated above: they are mostly based on dynamically allocated lists, sets and tables. For this type of applications, partly different methodologies and design tool support are required. This has been motivated by us in our global methodology proposal on the MATISSE approach [5]. At that time, we did not provide any design techniques or tool support yet. Recently, we have also proposed a more detailed methodology for the flow graph balancing step [17] but also in that paper no CAD techniques have been presented yet.

In this paper, we focus on the new techniques and tools in both our flow graph balancing and memory allocation/assignment steps, targeted to the communication network protocol domain. They will be explained in detail in Section 3.

Several problems in the state-of-the-art literature are related to these tasks. First, there is the register allocation domain which is fairly well understood by now. A nice literature overview of this domain can be found in [13]. The techniques used here start from a fully scheduled flow graph and are scalar-oriented. Many of these techniques construct a scalar conflict or compatibility graph and solve the problem using graph coloring or clique partitioning. This conflict graph is fully determined by the schedule. This means that no effort is spent in trying to come up with an optimal conflict graph.

In the less explored background memory allocation and assignment domain, the current techniques start from a given schedule [7], or perform first a bandwidth estimation step [1] which is a crude ordering that does not optimize the conflict graph either. These techniques have to operate on *groups of signals* instead of on scalars to keep the complexity acceptable, e.g. the *stream model* of Phideo [7] or the *basic sets* in the ATOMIUM approach [1].

In the scheduling domain, the techniques optimizing for the number of resources given the cycle budget are of interest to us. Also here most techniques operate on the scalar level, e.g. [10, 15]. The only exceptions currently are the Phideo stream scheduler [16] and the Notre-Dame rotation scheduler [11]. Many of these techniques try to reduce the memory related cost by estimating the required number of registers for a given schedule. Only few of them try to reduce the required memory bandwidth, which they do by minimizing the *number* of simultaneous memory accesses [15, 16]. They do not take into account *which* data is being accessed simultaneously. Also no real effort is spent to optimize the memory access conflict graphs such that subsequent register/memory allocation tasks can do a better job.

The main difference between our flow graph balancing and the related work discussed here is that we try to minimize the required memory bandwidth in advance by optimizing the access conflict graph for groups of scalars within a given cycle budget. We do this by putting ordering constraints on the flow graph, taking into account *which* memory accesses are being put in parallel (i.e., will show up as a conflict in the access conflict graph).

For the memory allocation and signal-to-memory assignment subtasks, our previous approach [1] is based on relatively greedy global optimization approaches which did not yet take into account memory access conflict constraints. As a result, the assignment technique could only work when the bandwidth constraints were not too tight and did not lead to conflict violations. In this paper, a novel method for signal-to-memory assignment is proposed which does incorporate a way to handle memory access conflict constraints. To achieve this, the original method had to be modified significantly. It also contains several improvements compared to other existing techniques because it takes into account the (extended) conflict graphs produced by the FG balancing step. Compared to other approaches, it also works on groups of scalars, thereby reducing the complexity significantly.

3 Exploration Environment

In this section our physical memory management script for network applications is explained. The input is a C or C++ description of the application, which is translated into a flow graph. The output is an optimized memory architecture. An overview of our exploration environment is shown in Fig. 1.

Given a flow graph of the application (cfr. upper left corner in Fig. 1), where for every memory access it is indicated which data is being accessed, an optimized memory architecture has to be derived during the allocation and assignment phase. However, because the memory architecture is derived before detailed scheduling, sufficient memory bandwidth must be foreseen such that the application can be scheduled within the given cycle budget afterwards. That is why the memory architecture is not derived directly from the flow graph. First a flow graph balancing step is performed, which tries to minimize the required memory bandwidth.

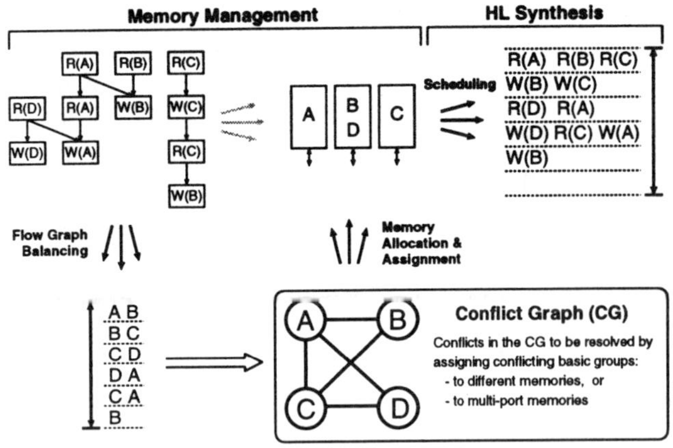

Figure 1: Physical Memory Management

Remark that due to the dynamic and data-dependent nature of the application domain, a number of steps of our High Level Memory Management (HLMM) script for Real-time Multi-dimensional Signal Processing (RMSP) applications [9] are not as useful in this context. In particular, the space reserved for dynamically allocated data of a certain type is assumed to be alive whole the time. This means that no in-place optimization [3, 4] can be done on it. Other very useful optimizations to lower the power consumption of RMSP applications, namely loop transformations to increase the locality of access and the exploitation of Data Reuse (leading to the introduction of memory hierarchy) [6], can also not be used in this context because no manifest loop nests are present and which data will be accessed is data-dependent and thus cannot be predicted.

In the next subsections, each subtask in our HLMM script for network applications will be explained.

3.1 Data Flow Analysis

The first task in the script translates the C/C++ description of the network component application into our flow graph format. It also performs data flow analysis to extract the parallelism from the sequential input description. For every data type in the application a virtual memory segment (VMS) is created. A VMS represents an amount of memory sufficient to store all objects of the corresponding data type. It also provides a dynamic memory management scheme to allocate (and deallocate) part of its memory space to data objects at run-time. For more details on how the dynamic allocated data is handled we refer to [5].

To increase the accuracy of the data flow analysis and to increase the freedom of subsequent memory management tasks, these virtual memory segments are partitioned further into non-overlapping *basic groups*. This is done in such a way that for every memory access, it is known at compile time which basic group is being accessed. For instance, assuming that for every memory access to a record, the field being accessed is known (a reasonable assumption in the application domain considered), all instances of a particular field (belonging to different objects of the same type) can be combined into one basic group. As a consequence the record layout plays an important role in the basic group partitioning. Indeed, fields grouped together in a common memory word, will be part of the same basic group. Currently our c2fg prototype tool is based on GNU's C compiler.

The basic groups are characterized by their bitwidth (extracted form the original specification), their number of words (extracted from the VMS definition), and the average number of memory accesses (derived from simulations). These figures are used for the area and power estimations.

3.2 Flow Graph Balancing for Memory Bandwidth Reduction

The flow graph balancing (FGB) step orders the memory accesses within the given cycle budget. In the context of ATM applications, nested loops almost never occur. The existing (single) loops are usually data-dependent (WHILE loop) spanning an entire task in a concurrent specification. In general, we can handle the memory management for the concurrent tasks separately, though also optimizations are feasible between tasks. The latter is however a topic of further research. So in this paper, we will assume that FGB has to operate on a single task, possibly generated after combining a number of initially concurrent tasks so that they are statically schedulable. Due to the very limited loop oriented characteristics, we can also restrict our current technique and prototype realization to handle a "flat" graph operating on the body of a single loop where typically complex dynamic data structures are accessed in a very complicated condition hierarchy.

Whenever two memory accesses to two basic groups (BGs) within the data structures occur in the same cycle, we say there is an access conflict between these two basic groups. All access conflicts are collected in a conflict graph, where the nodes represent basic groups, and the edges indicate a related conflict. These conflicts have to be resolved during the memory allocation and assignment steps. This can be done by assigning conflicting basic groups either to different memories or to a multiport memory. When all conflicts in the conflict graph are resolved during the memory assignment step, it is guaranteed that a valid schedule exists for the obtained memory architecture.

We have defined a cost function for these conflict graphs, such that more costly conflict graphs are likely to lead to more costly memory architectures. The cost function includes three weighted terms: 1) a measure for the minimum number of memories needed (the chromatic number[1] of the

[1] A *c-coloring* of a graph G is a partitioning of G's nodes in c partition

CG), 2) a term related to the number of conflicts in the conflict graph (each conflict is weighted with its importance), and 3) a term to minimize the number of self-conflicts (based on the notion of forces similar to IFDS [15]). For more details about the cost function we refer to [17]. The idea of flow graph balancing is then to come up with a partial ordering of the memory accesses that leads to a conflict graph with minimal cost.

When multiport memories are allowed in the memory architecture, more information is needed than present in a simple conflict graph. Therefore, we have defined an extended conflict graph (ECG). In an extended conflict graph, every conflict is annotated with the maximal number of simultaneous read accesses, the maximal number of simultaneous write accesses, and the maximal number of simultaneous memory accesses (i.e., read or write) that can occur between the conflicting basic groups during the execution of the algorithm. Also conflicts between more than two basic groups have to be taken into account, because several conflicting basic groups can be stored in a single multiport memory as long as the number of simultaneous memory accesses is not exceeding the access capabilities of the memory. This results in the inclusion of hyper edges in the ECG, indicating the conflicts between more than two basic groups. When multiport memories are available in the memory library, basic groups can be allowed to be in conflict with themselves, leading to self edges in the conflict graph. Obviously, such self conflicts will be very costly because they force the allocation of (expensive) multiport memories. Fig. 2 illustrates the difference between a conflict graph and an extended conflict graph for a given memory access ordering.

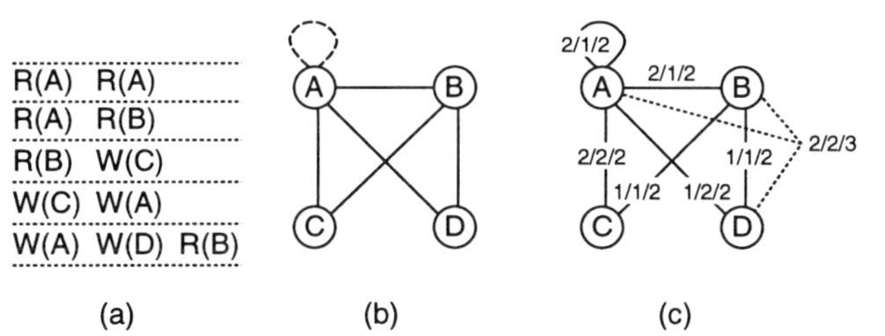

Figure 2: Extended Conflict Graph: (a) Schedule, (b) Conflict Graph, (c) Extended Conflict Graph. The R/W/RW numbers indicate the maximum number of simultaneous read operations (R), write operations (W), and read or write operations (RW) that occur for the given conflict.

Our current flow graph balancing tool uses an iterative search strategy to order the memory accesses similar to Improved Force Directed Scheduling of [15] but with totally different cost functions. As motivated above, no loops are supported as yet. That is a topic of current research. The prototype tool implements the principles introduced in [17].

classes $V = X_1+X_2+\cdots+X_c$ such that every two adjacent nodes belong to a different partition class. In this case, when the members of partition X_i are colored with color i, adjacent nodes will receive different colors. The chromatic number $\chi(G)$ is the smallest number c for which there exists a c-coloring of G.

3.3 Allocation and Assignment

The memory assignment technique discussed below is significantly different compared to existing techniques (including our own previous work [1]) because it has to take into account the (extended) conflict graphs produced by the FG balancing step. Compared to most other approaches, it also works on groups of scalars, thereby reducing the complexity significantly.

Once the (extended) conflict graph is available, we have all the inputs required to do a *valid and cost efficient* allocation and assignment. This is done by minimizing a cost function, containing weighted memory area and power terms, while taking into account all memory access constraints expressed by the conflict graph. The on-chip area and power models used in the cost function are proprietary from memory module vendors, so only relative figures are given in this paper.

During the *allocation* phase, the user decides on the *number of one-port memories* to be allocated. This should be at least the chromatic number of the conflict graph. Remark that this lower bound on the number of memories only holds for a library with one-port memories, which is always assumed further on in this paper[2]. The main reason for allocating more than the minimal number of memories is to reduce the power, as demonstrated further on. The practical upper bound on this number is the number of basic groups (cfr. Section 3.1).

During the *assignment* phase, each basic group is assigned *as a whole* to one of the allocated memories. This yields an assignment scheme. Usually a BG can only be assigned to *some* of the allocated memories, because of memory access conflicts with some of the basic groups assigned already. In order to find a minimum cost assignment scheme, we must explore the entire assignment search space, because classical global optimization approaches like (Mixed) Integer Linear Programming solvers do not work for our nonlinear problem. The assignment search space can be represented as a tree, as shown in Fig. 3. This tree has M^N leaves, where M denotes the number of memories allocated and N the number of basic groups to be assigned, so there is a huge amount of assignment possibilities. We use a *branch-and-bound* algorithm (called *B&B* in the sequel) with an effective bounding to search the complete assignment tree.

In our specific context, we have implemented a very effective B&B strategy which is different from our original approach [1] which did not incorporate the access conflicts.

[2] We also have a methodology for dealing with multi-port memories, but that feature is not implemented yet in our prototype software environment.

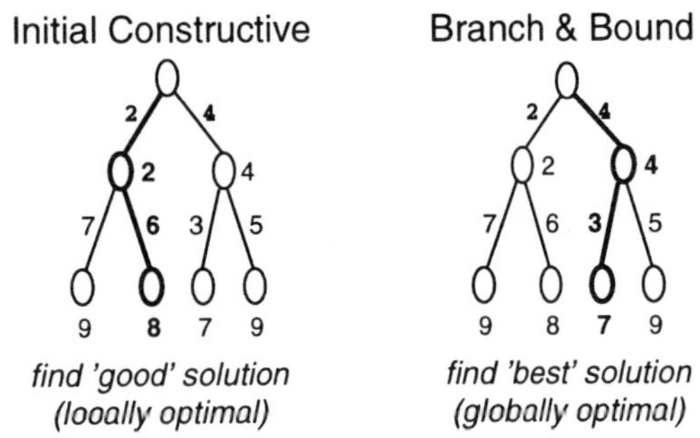

Figure 3: Different assignment algorithms for traversing the search tree: each tree level corresponds to the assignment of one basic group, each branch from a node corresponds to one memory.

Large parts of the tree can be cut away ('bound', i.e., pruning of subtrees) because of reasons, evaluated in this order:

1. paths which give rise to assignment schemes which are fully symmetric with already generated schemes can be discarded.
2. access conflicts between basic groups can be effectively checked due to the explicit ECG information, which allows to remove the corresponding sub trees
3. paths which have too high a cost from a certain basic group on can be pruned from that node. Currently we propose a simple but safe estimate of the minimum remaining cost which assumes storage in a common 1-port memory unit with minimum bitwidth for area and storage in separate 1-port memory units for power, and this for all remaining BGs.

Moreover, in order to have a low cost threshold (for pruning) at an early stage in the B&B process, we first perform an initial constructive assignment algorithm. This finds a local optimum in the search tree by iteratively assigning at each level in the search tree the corresponding basic group to the locally cheapest memory.

The order in which the basic groups are assigned also has an enormous effect on the run-time of the B&B algorithm. Experiments have shown that in most cases, ordering based on cost (i.e., assigning Basic Groups in the order from costly to cheap in terms of area/power consumption) leads to shorter run-times than ordering based on constraints (i.e., assigning BGs in the order from difficult-to-assign to easy-to-assign).

All the pruning criteria mentioned above have been implemented in our prototype tool. As a result, for the not very complex but still realistic example in Section 5, the CPU times can all be reduced below 1 minute on an HP 9000/715-50 workstation. Also for larger examples, experiments have confirmed that the run times remain very acceptable. Other performance improving measures have been developed by us and will be implemented in the future, such as splitting of the conflict graph into *(near-)independent clusters*, and additional cost pruning by *looking ahead* in the search tree.

4 Test Vehicle: STORM

As a test vehicle for our exploration experiments, we have selected part of the functional specification of an ASIC specified by Alcatel. The application is called STORM, which stands for ***S**TM1-level to **T**ransputer-netw**O**rk **R**elayer with **M**ultiple protocol handling*. It deals with protocol conversion between an ATM transportation layer and a transputer network (cfr. Fig. 4).

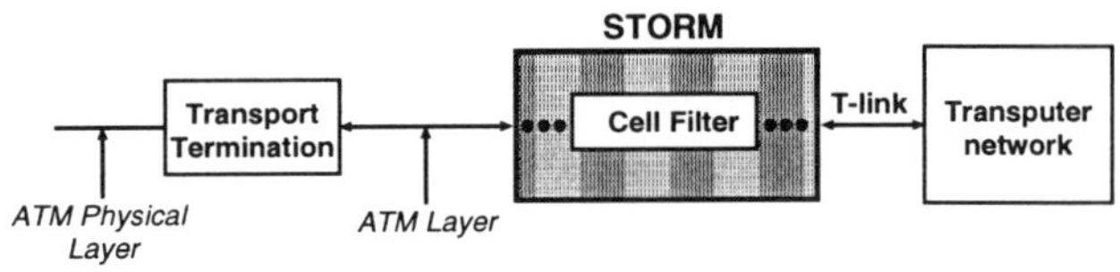

Figure 4: Position of STORM in ATM context

In this paper, we only focus on part of the STORM application, namely the cell filter. This part is representative, because, like the other subsystems, it is rather heavily IO-dominated: many background memory accesses have to be performed real-time within a limited number of cycles. It relates incoming ATM cells to allocated internal streams, by means of two search algorithms (cfr. Fig. 5), which can be combined in a pipelined way, i.e., they can be executed in parallel but acting on subsequent ATM cells.

Figure 5: Cell filter

Both search algorithms, binary search and 3-phase indirection, are looking for the same information but in a different way. The first algorithm uses the combination of some information fields of an ATM cell as one key into one big sorted table, stored in a *LookupRAM*. If the stream is not found in this table, the second algorithm looks for the stream in 3 steps. At each step, one of the subkeys (VPI, VCI, and MID extracted from the incoming ATM cell) is used as an index in the corresponding table, to find a pointer to the next table. The last table contains a pointer to the related stream.

5 Design Space Exploration: Results

In this section we show the results of our design space exploration experiments for the cell filter of the STORM application, by varying some of the design parameters. A similar exploration has also been done for the Segment Protocol Processor application from Alcatel [14], but due to space requirements these results are not presented in this paper.

5.1 Effect of Modified Record Layout

In network type applications, much of the data is stored in records. A record is a data structure that groups a number of related data items. These different data items are called the fields of the record. These fields can have different bitwidths. Several fields can be packed into a common word. How the different fields are packed into words is called the record layout.

The record layout heavily influences the optimal memory architecture, as it effects the number of memory accesses and the basic group partitioning. Here we present two extreme cases of record layout. In the first one, each field is stored in a separate memory word. The results are shown in Fig. 6. The access flow graph shows many memory accesses, because every field access requires its own memory access. There are many basic groups as can be seen in the conflict graph. The chromatic number of the conflict graph for a cycle budget of 6 cycles is 4. This means that in an architecture with only single port memories, at least 4 memory modules are needed to provide sufficient memory bandwidth.

A second case of record layout we have examined, is one where all fields of a given record are stored in a common memory word. The results for this are shown in Fig. 7. Now there are far less memory accesses, as can be seen in the access flow graph. Also the conflict graph is much simpler. In fact, in this case, there are no conflicts left in the graph which means that 1 memory is sufficient to meet the bandwidth requirements for a cycle budget of 6 cycles.

In between these two extreme cases, many other record layouts are possible. E.g., grouping the fields of a record such that they fit as good as possible into 32 bit words. This corresponds to the row labeled *32 bit* in Table 1.

The table entries contain relative area and power numbers for the total memory architecture resulting from allocation and assignment. The orders of magnitude are tens of mm^2

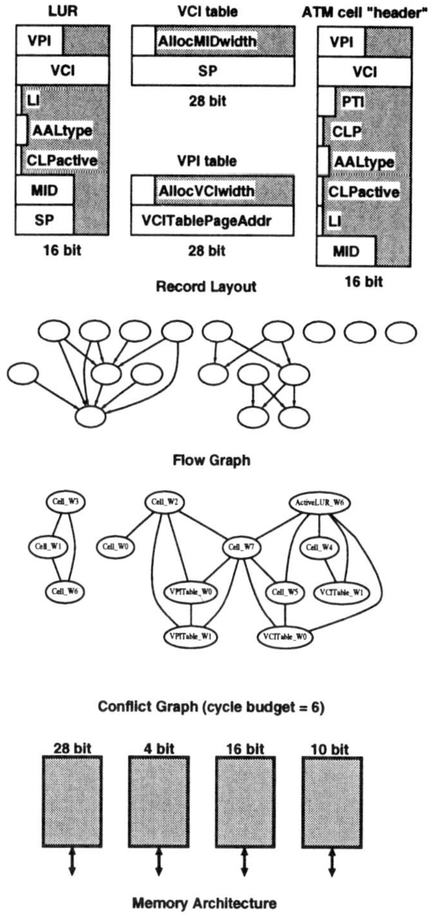

Figure 6: Record layout: 1 field per word

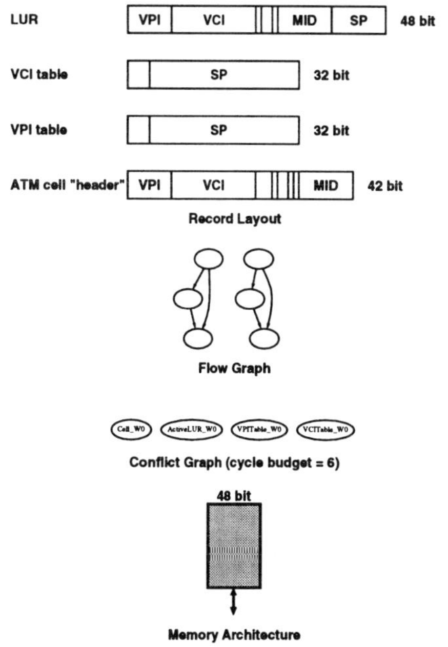

Figure 7: Record layout: 1 word per record

for area and a hundred mW for power. The area numbers are relative to the maximum area number of the table, whereas the power numbers are relative to the maximum power number.

	1 mem	2 mem	3 mem	4 mem
1 field 1 word	N/A	N/A	N/A	A: 1.00 P: 1.00
32 bit	N/A	A: 0.87 P: 0.96	A: 0.90 P: 0.60	A: 0.91 P: 0.50
1 word	A: 0.84 P: 0.92	A: 0.81 P: 0.64	A: 0.83 P: 0.39	A: 0.84 P: 0.34

Table 1: Optimal memory architecture for different record layouts and cycle budget = 6.

From this table we can conclude that, for this example, the area and power decrease when the records are packed into less words. Also fewer memories need to be allocated in case the records are packed into fewer words (the *N/A* entries denote that no valid assignment is possible for the corresponding number of allocated memories). Note also that the power decreases when more memories are allocated, while the area exhibits a minimum between the two extremes.

As indicated in subsection 3.1, it is also important to perform a partitioning of the original VMS data structures into basic groups. In this particular test-vehicle, this preprocessing results e.g. in a decrease of the number of self-conflicts (and hence minimal number of memory ports) from 4 to 2 for a cycle budget of 3 and the original record layout with 1 field stored per word.

5.2 Effect of the Cycle Budget

To see the effect of the cycle budget on the resulting memory architecture, we have varied the cycle budget from the critical path length (3 in the case of one field per word) to the number of memory accesses in the flow graph of the application (18 in this case). Assuming that every memory access requires exactly one cycle[3], the minimal bandwidth is reached (i.e., 1 memory port) for this amount of cycles. Allocating more cycles will have no further effect on the memory architecture. The resulting chromatic number is shown for each cycle budget in Fig. 8. Remark that in general this value decreases as more cycles are available. However, this is not always the case (e.g., cycle budget = 11) because during flow graph balancing also other cost terms than the chromatic number are taken into account (e.g., the number of conflicts, each weighed with their cost). Ideally the total cost should be decreasing monotonically (as is the case in Fig. 8). This is not guaranteed, however, with our FGB tool as it is a heuristic that does not necessarily find the global optimum. The CPU times for these experiments are all below 20 seconds on an HP 9000/715-50 workstation.

[3]This is usually the case for SRAM memories. If different types of RAMs are used, appropriate IO profiles (with larger latency values for instance) have to be used.

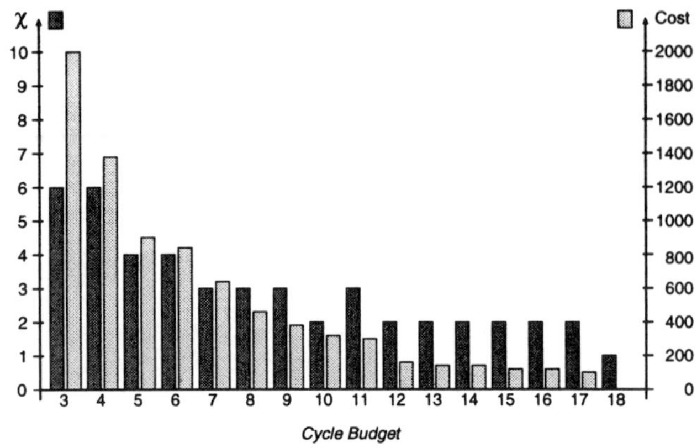

Figure 8: Chromatic number and FGB cost function for different cycle budgets.

5.3 Effect of Allocation/Assignment

To see the effect of the *number of allocated memory modules*, we have varied that number for a cycle budget equal to 6 and a record layout with 1 field per word. The results are shown in Table 2.

The minimal required number of 1 port memories is 4 (i.e., the chromatic number of the conflict graph for this cycle budget). The maximal number of memories equals the number of basic groups, because basic groups are assigned as a whole to a single memory. The values in the table were optimized both for area and power with equal weight factors. As can be seen in the table, the area decreases with increasing number of memories, up to a certain number of memories. The reason for this is that, because of the possibly different bit widths of the allocated memories, less bits have to be wasted when more memories are available in the memory architecture. Therefore basic groups with different bitwidths have a larger chance to be assigned to different memories, which will then have a bitwidth adapted to them.

However, from a certain number of memories on, the area starts increasing again with increasing number of memories, because more memories involves more overhead. The power keeps decreasing when more memories are allocated. Remark also that the optimal solution (w.r.t. weighted area/power cost) for these cost parameters is an architecture with 11 memories. Therefore, even when more than 11 memories are allocated, only 11 of them will be effectively used. Note, that the cost of interconnect is not yet taken into account in our cost functions. When this is done, the optimal values will probably occur for a smaller number of memories.

#Mem	Area (rel)	Power (rel)	Weighted Cost
4	1.00	1.00	1.00
5	0.93	0.76	0.87
6	0.92	0.70	0.85
7	0.92	0.63	0.82
8	0.94	0.54	0.80
9	0.94	0.50	0.79
10	0.95	0.48	0.79
11	0.95	0.47	0.78
12	0.95	0.47	0.78
13	0.95	0.47	0.78

Table 2: Optimal cost results for different 1-port memory allocations for a cycle budget of 6.

The resulting basic group to memory assignment depends upon the exploration of the assignment search space, i.e., on the assignment algorithm used. We mainly used the *initial constructive* and the *branch-and-bound* assignment algorithms, as shown in Fig. 3. Experiments with both alternatives revealed for instance that, for the parameters mentioned above and 5 memories allocated, area and power for initial constructive are 10% resp. 30% higher than the area and power obtained with B&B. This shows that performing an extensive assignment search space exploration, as is done by the B&B algorithm, can reduce the area to some extent, and can have a significant effect on the power consumption of the final solution.

6 Conclusions

We have shown that significant better results than conventional design approaches can be obtained for the memory organization of network protocol applications when the full search space is well explored. The methodology and prototype tools presented in this paper allow for the first time to do this in a fast and thorough way starting from a high abstraction level. The run-times of our tools are all below 1 minute on an HP 9000/715-50 workstation for the application presented. In this way the system design time for these applications is significantly reduced while the final power and area cost are improved at the same time. This is of crucial importance for cost effective designs in the telecom network protocol industry

Acknowledgments: We gratefully acknowledge the discussions with our colleagues at IMEC and Alcatel and especially the contributions of M.Miranda, M.Genoe, M.Eyckmans, and P.Six.

References

[1] F.Balasa, F.Catthoor, H.De Man, "Background memory area estimation for multi-dimensional signal processing systems", *IEEE Trans. on VLSI Systems*, vol. 3, no. 2, pp. 157-172, June 1995.

[2] R.Camposano, W.Wolf (eds.), "Trends in high-level synthesis", Kluwer Academic Publishers, Boston, 1991.

[3] E.De Greef, F.Catthoor, H.De Man, "Memory Size Reduction through Storage Order Optimization for Embedded Parallel Multimedia Applications", *Intnl. Parallel Proc. Symp.(IPPS) in Proc. Wsh on "Parallel Processing and Multimedia"*, Geneva, Switzerland, April 1997.

[4] E.De Greef, F.Catthoor, H.De Man, "Array Placement for Storage Size Reduction in Embedded Multimedia Systems", *Intnl. Conf. on Applic.-Spec. Array Processors (ASAP)*, Zurich, Switzerland, July 1997.

[5] G.de Jong, B.Lin, C.Verdonck, S.Wuytack, F.Catthoor, "Background memory management for dynamic data structure intensive processing systems", *Proc. IEEE Int. Conf. Comp. Aided Design*, San Jose CA, pp.515-520, Nov. 1995.

[6] J.P.Diguet, S.Wuytack, F.Catthoor, H.De Man, "Formalized Methodology for Data Reuse Exploration in Hierarchical Memory Mappings", accepted for *Proceedings of IEEE International Symposium on Low Power Electronics and Design*, Monterey CA, Aug. 1997.

[7] P.Lippens, J.van Meerbergen, W.Verhaegh, A.van der Werf, "Allocation of Multiport Memories for Hierarchical Data Streams", *Proc. IEEE Int. Conf. Comp.-Aided Design*, pp. 728-735, Santa Clara, Nov. 1993.

[8] T.Meng, B.Gordon, E.Tsern, A.Hung, "Portable video-on-demand in wireless communication", special issue on "Low power design" of the *Proc. of the IEEE*, Vol. 83, No. 4, pp. 659-680, Apr. 1995.

[9] L.Nachtergaele, F.Catthoor, F.Balasa, F.Franssen, E.De Greef, H.Samsom, H.De Man, "Optimization of memory organization and partitioning for decreased size and power in video and image processing systems", *IEEE Int'l Workshop on Memory Technology, Design and Testing*, pp. 82-87, San Jose CA, Aug. 1995.

[10] P.Paulin, J.Knight, "Force-directed scheduling for the behavioral synthesis of ASIC's", *IEEE Trans. on CAD*, Vol. 8, No. 6, pp. 661-679, June 1989.

[11] N.Passos, E.Sha, "Push-up scheduling: optimal polynomial-time resource constrained scheduling for multi-dimensional applications", *Proc. IEEE Int. Conf. Comp. Aided Design*, San Jose CA, pp.588-591, Nov. 1995.

[12] L.Ramachandran, D.Gajski, V.Chaiyakul, "An algorithm for array variable clustering", *Proc. European Design and Test Conf.*, pp. 262-266, Paris, Mar. 1994.

[13] L.Stok, "Data path synthesis", *INTEGRATION, the VLSI journal*, Vol 18, pp . 1-71, June 1994.

[14] Y.Therasse, G.H.Petit, M.Delvaux, "VLSI architecture of a SMDS/ATM router", *Annales des Télécommunications*, 48, no 3-4, pp.166-180, 1993.

[15] W.Verhaegh, P.Lippens, E.Aarts, J.Korst, J.van Meerbergen, A.van der Werf, "Improved Force-Directed Scheduling in High-Throughput Digital Signal Processing", *IEEE Transactions on CAD and Systems*, Vol. 14, no 8, Aug. 1995.

[16] W.Verhaegh, P.Lippens, E.Aarts, J.van Meerbergen, A.van der Werf, "Multi-dimensional periodic scheduling: model and complexity", *Proc. EuroPar Conference*, Lyon, France, August 1996. "Lecture notes in computer science" series, Springer Verlag, pp.226-235, 1996.

[17] S.Wuytack, F.Catthoor, G.de Jong, B.Lin, H.De Man, "Flow Graph Balancing for Minimizing the Required Memory Bandwidth", *Proc. 9th ACM/IEEE Intnl. Symp. on System-Level Synthesis*, La Jolla CA, pp.127-132, Nov. 1996.

Optimization of the Background Memory Utilization by Partitioning

Uwe Eckhardt

IEE, TU Dresden
eckhardt@iee.et.tu-dresden.de

Renate Merker

IEE, TU Dresden
merker@iee1.et.tu-dresden.de

Abstract

The skilful utilization of the memory structure of a processor and of its background memory may crucially affect the system performance. We propose a restructuring of for-loop programs by hierarchical partitioning which improves the properties of the algorithm with respect to the memory utilization. We consider the problem for regularly connected processor arrays (where single processors are a special case) and for a memory structure which is subdivided into local foreground memory (register) and background memory with up to three levels (cache, RAM, mass storage). The extension of the lifetime of a variable on an inner memory level, i.e. the decrease of the number of read accesses to more outer memory levels is the object of the proposed method.

1 Introduction

The efficient processing of for-loop programs on single processors and processor arrays is a well studied problem, which incorporates program restructuring techniques and high level synthesis methods. There exist compiler systems for the mapping of for-loop programs, where indexing functions of the variables are affine functions of the loop indices, onto (piecewise) regularly connected processor arrays. An essential step in these compilers is the transformation of global data dependencies into local data dependencies by localization procedures [16]. The resulting algorithms, the so-called (conditional) uniform recurrence equations (URE's) [8], are considered in this paper.
The loop-transformations splitting and interchange have been discussed in [18, 17] for the memory synthesis for these algorithms. We apply these ideas in a systematic partitioning framework. We extend the improvement of the foreground memory utilization by an improvement of the utilization of different levels of the background memory. The background memory is supposed to be structured in a cache level, a RAM level and a mass storage level (see Fig.(1)). The extension of the lifetime of a variable on an inner memory level , i.e. the decrease of the number of read accesses to more outer memory levels is the object of the proposed method. The foreground memory is supposed to be local memory (register) associated to each processor and the background memory belongs to the peripheral system.
We propose a hierarchical application of the well studied locally parallel, globally sequential (LPGS)- and locally sequential, globally parallel (LSGP)-partitioning schemes [7, 9, 11, 10, 12, 6, 1, 14, 15] for the systematic solution of the memory utilization problem.
The remainder of the paper is organized as follows: Section 2 presents the key-concept of the optimization of the background memory utilization by hierarchical partitioning for the example of the matrix multiplication. Section 3 is devoted to the mathematical model of the proposed method. Finally, a conclusion is given.

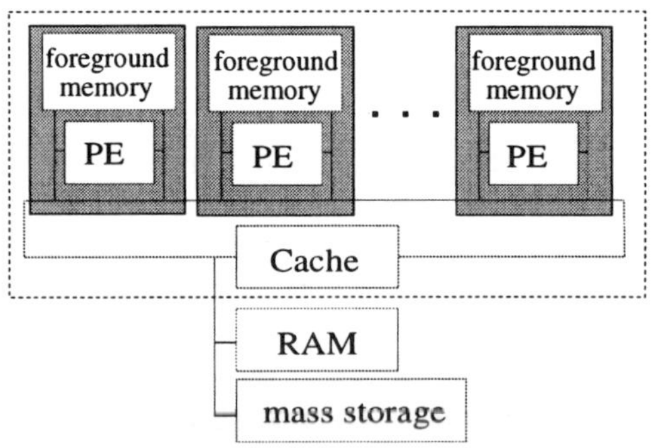

Figure 1: *Memory structure*

2 The key-concept of hierachical partitioning

In this section we illustrate the key-concept of the hierarchical partitioning for the example of the matrix multiplication $\mathbf{C} = \mathbf{QW}$ of two square matrices of

1080-1820/97 $10.00 © 1997 IEEE

order N.

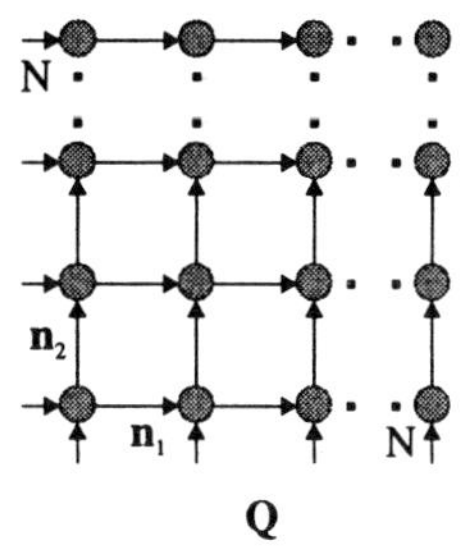

Figure 2: *Example: matrix muliplication array*

2.1 Foreground memory utilization and I/O access behaviour to the background memory

The first step of the mapping of the matrix multiplication onto the processor array results in a so-called full size array (see section 3.1). Here, the extend (i.e. the number of processing elements) of the array depends on the order of the matrices. The full size array for the considered matrix multiplication is shown in Fig.(2). The variables w and q are propagated along the directions $\mathbf{n}_1$ and $\mathbf{n}_2$ respectively. The variables c are fixed on the processing elements.

The goal is the processing of the algorithm on a structure with a defined number of processing elements and a defined I/O-behaviour between the processing elements and between the whole array and the peripheral system. Hence, partitioning techniques have to be applied. Suppose as target architecture a processor array with two times two processing elements and with given I/O-capacities of the interconnections between the processing elements of the array and between the array and the peripheral system, which contains the background memory. The co-partitioning method [4] which represents a simultaneous LPGS- and LSGP-partitioning (see section 3.2) has to be applied to adjust the full size array Fig.(2) to the given constraints. Fig.(3) depicts the co-partitioning of the full size array. The small rectangles symbolize the LSGP-partitions. All tasks belonging to an LSGP-partition are performed sequentially on one processing element of the partition. Thus, there remains only one processing element for each LSGP-partition. The remaining processing elements are drawn black. The doted lines within the LSGP-partitions in Fig.(3) represent the introduced sequential scheduling. All LSGP-partitions are active in parallel. Hence, the $N \times N$ full size array

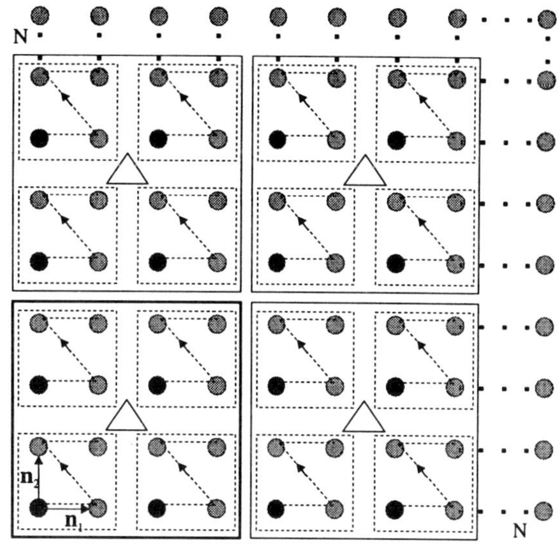

Figure 3: *Foreground memory allocation by two-fold partitioning*

has been transformed into an LSGP-partitioned array of $\frac{N}{2} \times \frac{N}{2}$ processing elements. The large rectangles in Fig.(3) symbolize the LPGS-partitions. The processors of an LPGS-partition are active in parallel and the partitions are scheduled sequentially. Each partition consists of two times two processing elements.

The LSGP-part of the co-partitioning defines the size of the local foreground memory, which has to be added to each processing element. The size of this local memory depends directly on the number of processing elements, which belong to the LSGP-partition. The LSGP-partitioning and hence, the addition of foreground memory leads to a decrease of the I/O-demand between the processing elements of the array and between the whole array and the peripheral system. A partition of two times two processing elements has to receive two variables w and two variables q and has to put out also two variables w and two variables q for each processing cycle. Hence, a partition of two times two processing elements has an I/O-demand with the peripheral system of eight data items in each processing cycle. Each LPGS-partition in the co-partitioned array shown in Fig.(3) has an I/O-demand of 16 data items. However, this I/O-demand is distributed over the four sequential cycles introduced by the LSGP-partitioning. Hence, we achieve a mean value of the I/O-demand of four data items per clock cycle.

Therefore, we can realize a balancing of the size of the foreground memory and the I/O-access behaviour of the array to the background memory by co-partitioning. The co-partitioning represents a partitioning with a hierarchy of two levels (see section 3.2).

2.2 Improvement of the cache utilization

Suppose, that a co-partitioning has been applied to achieve an array of two times two processors with an I/O-behaviour which is adjusted to the I/O-constraints imposed by the peripheral system. The resulting LPGS-partitions (co-partitions) have to be scheduled sequentially. The elements symbolized by the triangles in Fig.(4) represent the LPGS-partitions (co-partitions) achieved in the previous foreground memory allocation process. The sequential scheduling of the LPGS-partitions can be given by two nested for-loops, where one loop is associated with direction $\mathbf{n}_1$ and the other with $\mathbf{n}_2$. The loop nest can be arbitrarily selected.

We assume, that we have a cache of $4 * 10^3$ words

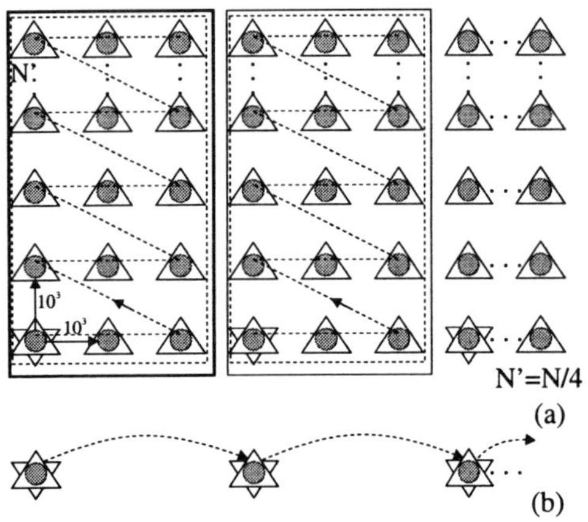

Figure 4: *Improved cache level utilization by four-fold partitioning*

and an I/O-demand of 10^3 words in direction $\mathbf{n}_1$ and $\mathbf{n}_2$ (i.e. 10^3 words have to be transferred to the array in each direction $\mathbf{n}_{1,2}$ for the processing of all tasks which are allocated to an LPGS-partition (co-partition)). We could store the first $3 * 10^3$ words of variables q and the 10^3 words of variable w in the cache in the case of a scheduling of the LPGS-partitions first along direction $\mathbf{n}_1$ (inner loop) and then along direction $\mathbf{n}_2$ (outer loop). Hence, all variables w and the first $3 * 10^3$ words of variables q have to be read once and the remaining $(N' - 3) * 10^3$ words of variables q have to be read N' times from the RAM or mass storage level. We obtain for this scheduling a read access of $(N'^2 - 2N' + 3) * 10^3$ words to the RAM and mass storage level. The partitioning and scheduling like shown in Fig.(4) leads to a decrease of this access to $(\frac{N'^2}{3} + N') * 10^3$ words. Hence, we obtain for large N' a decrease of factor three. The doted lines in Fig.(4) represent the sequential scheduling within the partitions and of the partitions respectively.
Obviously, this partitioning represents also a co-partitioning, where the LPGS-partitions consist of one processing element, which is symbolized by the white star in Fig.(4). Fig.(4b) illustrates the sequential scheduling of the LPGS-partitions of the second co-partitioning. Hence, we can apply the same formal framework like for the foreground memory allocation for the representation of this partitioning. Therefore, we have to apply a partitioning with a hierarchy of four levels for foreground memory allocation and improvement of the cache utilization (see section 3.4). The access to the RAM and mass storage occurs on the boundary of the partitions. Hence, the optimization of the cache utilization is given by the determination of LSGP-partitions (in the second co-partitioning) of maximal size, which can realize the internal communication (data transfers within the partition) via the cache, i.e. without access to the RAM or mass storage.

read access to level	improved cache	improved cache & RAM
RAM & MS	$\frac{N'^2}{3} + N'$	$\frac{N'^2}{3} + 2N'$
MS	$\frac{N'^2}{6} + 1.5N'$	$3N'$

Table 1: Access to RAM and mass storage (MS) level

2.3 Improvement of the RAM utilization

Suppose a RAM of $N'/2 * 10^3$ words in the memory structure, which is attached to the array. If we apply the partitioning as shown in Fig.(4), then we can store $N'/2 * 10^3$ words w which have to be transferred from the left to right side of the array (i.e. along $\mathbf{n}_1$) in the RAM for the reuse in the next partition. The remaining $N'/2 * 10^3$ words which have to be transferred along $\mathbf{n}_1$ have to be read from the mass storage each time a new partition is scheduled. Therefore, the total amount of read accesses to the mass storage during the processing of the matrix multiplication is $(\frac{N'^2}{6} + 1.5N') * 10^3$. A decrease of the amount of read accesses to the mass storage can be achieved as follows: We modify the extend of the partitions shown in Fig.(4) in direction $\mathbf{n}_2$ from N' to one. Thereafter, we perform a third co-partitioning analogous to the second one. Fig.(5a) depicts the second co-partitioning after the third one has been performed, where the white stars represent the LPGS-partitions. The extend of the partitions in direction $\mathbf{n}_2$ has been changed from one to $N'/2$. Fig.(5b) illustrates the third co-partitioning. Thereby, we achieved a decrease of the amount of read accesses to the mass storage from $(\frac{N'^2}{6}+1.5N')*10^3$ to $3N'*10^3$. However, the amount of read access to the RAM and mass storage has been increased from $(\frac{N'^2}{3}+N')*10^3$ to $(\frac{N'^2}{3}+2N')*10^3$ in comparison with the previous result shown in Fig.(4). The optimization of the RAM utilization is given by

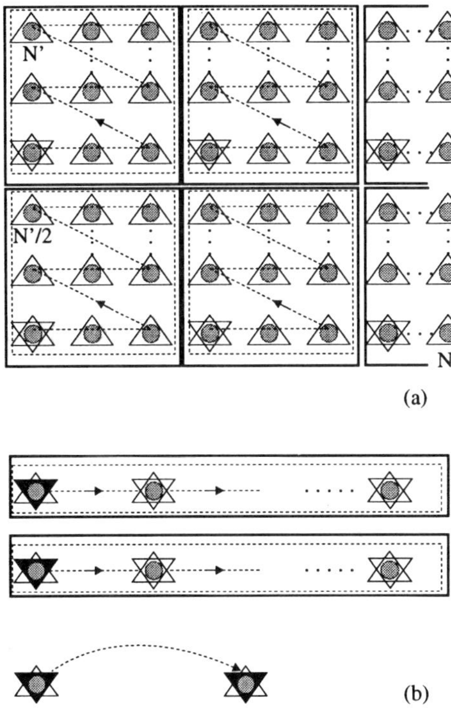

Figure 5: *Improved RAM level utilization by six-fold partitioning*

the determination of LSGP-partitions (in the third co-partitioning) of maximal size, which can realize the internal communication (data transfers within the partition) via the RAM, i.e. without access to the mass storage. Hence, we obtain a partitioning with a hierarchy of six levels (see section 3.5).
The achieved results are recapitulated in Tab.(1).

3 Formalized partitioning

3.1 Basic definitions

We introduce the definitions of the considered system of uniform recurrence equations (URE's), of the mapping onto systolic arrays and of the partitioning.

Definition 1 (System of URE's) *A system of uniform recurrence equations (URE's) on a linearly bounded lattice is a set of equations*

$$y_j[\mathbf{i}] = F(x_1[\mathbf{i}-\mathbf{d}_{x,1}],\ldots,x_l[\mathbf{i}-\mathbf{d}_{x,l}],$$
$$y_1[\mathbf{i}-\mathbf{d}_{y,1}],\ldots,y_m[\mathbf{i}-\mathbf{d}_{y,m}]),\, j=1,\ldots,m \qquad (1)$$

where the index space $\mathcal{I}$ of these equations is a linearly bounded lattice [1]

$$\mathcal{I} = \{\mathbf{i} \mid \mathbf{i} = \mathbf{Bz}+\mathbf{b}_0,\ \mathbf{Ai} \geq \mathbf{a}_0\}, \qquad (2)$$

$\mathbf{i} \in \mathcal{I}$ are index points and $\mathbf{i},\mathbf{z},\mathbf{b_0} \in \mathcal{Z}^n, \mathbf{a_0} \in \mathcal{Q}^m, \mathbf{B} \in \mathcal{Z}^{n\times n}, rank\mathbf{B} = n, \mathbf{A} \in \mathcal{Q}^{m\times n}$. The data dependence vectors $\mathbf{d}_{.,.} \in \mathcal{D}$ are constant, where $\mathcal{D}$ is the set of all data dependence vectors in (1).

Example: The multiplication $\mathbf{C} = \mathbf{QW}$ of two square matrices of order N is given by:

$$c[\mathbf{i}] = c[\mathbf{i}-(0\ 0\ 1)^t] + q[\mathbf{i}-(0\ 1\ 0)^t] * w[\mathbf{i}-(1\ 0\ 0)^t]$$
$$\mathbf{i} \in \mathcal{I} = \{\mathbf{i} = (i\ j\ k)^t \mid \mathbf{i} = \mathbf{Iz},$$
$$\begin{pmatrix} \mathbf{I} \\ -\mathbf{I} \end{pmatrix} \mathbf{i} \geq (0\ 0\ 0\ (1-N)\ (1-N)\ (1-N))^t\},$$

where $\mathbf{I}$ is the identity matrix of order three. The data dependence vectors are $\mathbf{d}_c = (0\ 0\ 1)^t$, $\mathbf{d}_q = (0\ 1\ 0)^t$, and $\mathbf{d}_w = (1\ 0\ 0)^t$. •

The scheduling of the computation tasks $\mathbf{i} \in \mathcal{I}$ in an array is defined by scalar product with the scheduling vector $\boldsymbol{\tau}$. The allocation of the computation tasks $\mathbf{i} \in \mathcal{I}$ to processors of the array is defined by a linear projection with a projection vector $\mathbf{u}$.

Definition 2 (Mapping) *The mapping M of the algorithm onto the systolic array is of the following form:*

$$M : \mathcal{I} \to \mathcal{J} \subseteq \mathcal{T}\times\mathcal{P},\ M : \mathcal{D} \to \mathcal{C} \subseteq \mathcal{T}_v \times \mathcal{V}, \qquad (3)$$

where $\mathcal{J}$ is the time processor space, $\mathcal{T} = \{t \mid t = \boldsymbol{\tau}\mathbf{i}\}$ is the set of time instances at which the iterations $\mathbf{i}$ are processed, and $\mathcal{P} = \{\mathbf{p} \mid \mathbf{p} = \mathbf{Si}\}$ is the processor space, where $\mathbf{Su} = \mathbf{0}, \mathbf{S} \in \mathcal{Z}^{n-1\times n}, rank\ \mathbf{S} = n-1$. $\mathcal{C}$ is the set of interconnection primitives, where $\mathcal{V} = \{\mathbf{v} \mid \mathbf{v} = \mathbf{Sd}\}$ is the set of interconnections which have to be realized in the array, and $\mathcal{T}_v = \{t_v \mid t_v = \boldsymbol{\tau}\mathbf{d}\}$ is the set of time delays associated with the interconnections. The constraints $\forall \mathbf{d} \in \mathcal{D} : \boldsymbol{\tau}\mathbf{d} > 0$ (causality) and $\boldsymbol{\tau}\mathbf{u} \neq 0$ (conflict free) have to be satisfied. The resulting array architecture is called full size array.

[1]This definition represents a special case of the more general definition of linearly bounded lattices in [13] for a full rank matrix B.

We claim, that $|\boldsymbol{\tau}\mathbf{u}|/gcd(\boldsymbol{\tau}\mathbf{B}) = 1$ holds, which is equivalent to the fact that a processing element is active in each clock cycle. The processor space is a linearly bounded lattice.

Example: We achieve the matrix multiplication array shown in Fig.(2), if we apply the scheduling vector $\boldsymbol{\tau} = (1\ 1\ 1)^t$ and a projection vector $\mathbf{u} = (0\ 0\ 1)^t$. The processor space $\mathcal{P}$ is given by

$$\mathbf{p} \in \mathcal{P} = \{\mathbf{p} = (i\ j)^t \mid \mathbf{p} = \mathbf{I}\mathbf{z},$$

$$\begin{pmatrix} \mathbf{I} \\ -\mathbf{I} \end{pmatrix} \mathbf{p} \geq (0\ 0\ (1-N)\ (1-N))^t\},$$

where $\mathbf{I}$ is the identity matrix of order two. The resulting interconnections in the array are $\mathbf{v}_c = (0\ 0)^t$, $\mathbf{v}_q = (0\ 1)^t$, and $\mathbf{v}_w = (1\ 0)^t$. •

In general, constraints in the structure of the array (e.g. number of processing elements) and I/O-demand cannot be taken into account with this mapping procedure. Hence, partitioning techniques have to be applied. In the following we define the subdivision of the algorithm and the array into congruent partitions.

Definition 3 (Partitioning) *The partitioning of the index space $\mathcal{I}$ of a system of URE's is given by the partitions $\mathcal{I}_\pi(\mathbf{i}')$ in the index space together with the set $\mathcal{I}'_\pi$ as follows:*

$$\mathcal{I}_\pi(\mathbf{i}') = \{\mathbf{i} | \mathbf{i} = \mathbf{i}' + \mathbf{G}^*\boldsymbol{\kappa}^*,\ \mathbf{i} \in \mathcal{I},\ 0 \leq \kappa_z < \vartheta_z,$$

$$z = 1, \ldots, n-1\}, \tag{4}$$

where $\kappa_z, \vartheta_z \in \mathcal{Z}, \boldsymbol{\kappa}^* \in \mathcal{Z}^n, \mathbf{G}^* = [\mathbf{G}, \mathbf{u}] = [\mathbf{g}_1 \ldots \mathbf{g}_{n-1}\,\mathbf{u}]$. $\mathbf{G}$ is an integral basis of the hyperplane $\boldsymbol{\tau}\mathbf{i} = 0$, and $\mathbf{i}' = \mathbf{G}\boldsymbol{\kappa}, \boldsymbol{\kappa} \in \mathcal{Z}^{n-1}$. The set $\mathcal{I}'_\pi$ of (reference) vectors $\mathbf{i}'$ is given by

$$\mathcal{I}'_\pi = \{\mathbf{i}' | \mathbf{i}' = \mathbf{i}_{zero} + \mathbf{G}\boldsymbol{\Theta}\mathbf{c},\ \mathbf{L}\mathbf{c} \geq \mathbf{l},\ \mathbf{c} \in \mathcal{Z}^{n-1},$$

$$\mathbf{i}_{zero} = \mathbf{G}\boldsymbol{\kappa},\ \boldsymbol{\kappa} \in \mathcal{Z}^{n-1}\} \tag{5}$$

with $\boldsymbol{\Theta} = diag(\vartheta_i)$, and the polyhedron $\mathbf{L}\mathbf{c} \geq \mathbf{l}$ is given by the successive projection of the convex hull of

$$\begin{pmatrix} \mathbf{A}\mathbf{G}^* & \mathbf{A}\mathbf{G}\boldsymbol{\Theta} \\ \mathbf{I} & \\ -\mathbf{I} & 0 \end{pmatrix} \begin{pmatrix} \kappa_1 \\ \vdots \\ \kappa_n \\ \mathbf{c} \end{pmatrix} \geq \begin{pmatrix} \mathbf{a}_0 - \mathbf{A}\mathbf{i}_{zero} \\ 0 \\ -(\boldsymbol{\vartheta} - \mathbf{1}) \end{pmatrix} \tag{6}$$

along the κ_j-axes, $j = 1, \ldots, n$, i.e. by Fourier-Motzkin elimination, where $\boldsymbol{\vartheta} - \mathbf{1} = (\vartheta_1 - 1 \ldots \vartheta_{n-1} - 1)^t$, and $\mathbf{I} \in \mathcal{Z}^{(n-1)\times(n-1)}$ is the identity matrix.

Hence, the processor space $\mathcal{P}$ is subdivided into partitions

$$\mathcal{P}_\pi(\mathbf{p}') = \{\mathbf{p} | \mathbf{p} = \mathbf{p}' + \mathbf{N}\boldsymbol{\kappa},\ \mathbf{p} \in \mathcal{P},\ 0 \leq \kappa_z < \vartheta_z,$$

$$z = 1, \ldots, n-1\} \tag{7}$$

with $\mathbf{N} = [\mathbf{n}_1 \ldots \mathbf{n}_{n-1}]$, $\mathbf{n}_j = \mathbf{S}\mathbf{g}_j$, and $\mathbf{p}' \in \mathcal{P}'_\pi$,

$$\mathcal{P}'_\pi = \{\mathbf{p}' | \mathbf{p}' = \mathbf{S}\mathbf{i}_{zero} + \mathbf{N}\boldsymbol{\Theta}\mathbf{c},\ \mathbf{L}\mathbf{c} \geq \mathbf{l}\}. \tag{8}$$

Example: Let us consider the partitioning of the processor space symbolized by the small rectangles in Fig.(3). The parameters are $\mathbf{N} = (\mathbf{n}_1\ \mathbf{n}_2)$ and $\vartheta_1 = \vartheta_2 = 2$, $\boldsymbol{\Theta} = diag(\vartheta_i)$. •

3.2 Foreground memory utilization

A two-fold partitioning, called co-partitioning [4], is necessary for the scaling of the structure of the array and for the balancing of the size of the foreground memory and of the I/O-access behaviour of the array to the peripheral system (background memory). The co-partitioning is a simultaneous LPGS- and LSGP-partitioning. The LPGS-partitioning leads to an array of fixed size, which is independent of the algorithm's size. However, for a fixed number of processing elements exists a minimum for the volume of data which has to be exchanged in one clock cycle with the peripheral system. If this minimum exceeds the I/O-capacity, then wait cycles have to be introduced in the case of a pure LPGS-partitioning. An LSGP-partitioning of the LPGS-partitioned array serves in the co-partitioning for the elimination of this effect. The interesting property of the LSGP-scheme is that it allows a tuning of the I/O-demand by tuning the size of the partitions. The co-partitioning is defined as follows:

Definition 4 (Co-partitioning)
The co-partitioning of the algorithm's index space $\mathcal{I}$ of an system of URE's is given by the following co-partitions in the index space:

$$\mathcal{I}_\pi(\mathbf{i}') = \{\mathbf{i} | \mathbf{i} = \mathbf{i}' + \kappa_n\mathbf{u} + \mathbf{G}(\boldsymbol{\Theta}^{LS}\boldsymbol{\kappa}^{LP} + \boldsymbol{\kappa}^{LS}),$$

$$\mathbf{i} \in \mathcal{I},\ \mathbf{i}' \in \mathcal{I}'_\pi\}, \tag{9}$$

where $\boldsymbol{\Theta}^{LS} = diag(\vartheta_j^{LS})$, $0 \leq \kappa_j^{LS} < \vartheta_j^{LS}$, $0 \leq \kappa_j^{LP} < \vartheta_j^{LP}, j = 1, \ldots, n-1$, and by the set $\mathcal{I}'_\pi = \{\mathbf{i}'\}$, which is defined according to equations (5) and (6), where $\boldsymbol{\Theta} = \boldsymbol{\Theta}^{LP}\boldsymbol{\Theta}^{LS}$.
The co-partitioning of the processor space $\mathcal{P}$ is given by the co-partitions in the processor space as follows:

$$\mathcal{P}_\pi(\mathbf{p}') = \{\mathbf{p} | \mathbf{p} = \mathbf{p}' + \mathbf{N}(\boldsymbol{\Theta}^{LS}\boldsymbol{\kappa}^{LP} + \boldsymbol{\kappa}^{LS}),$$

$$\mathbf{p} \in \mathcal{P},\ \mathbf{p}' \in \mathcal{P}'_\pi\}, \tag{10}$$

and by the set $\mathcal{P}'_\pi = \{\mathbf{p}'\}$, which is defined according to equation (8), where $\boldsymbol{\Theta} = \boldsymbol{\Theta}^{LP}\boldsymbol{\Theta}^{LS}$. We apply here the abbreviations LP instead of LPGS and LS instead of LSGP.

The structure of the co-partitioned array is given by the LPGS-parameters. The transformation of the internal interconnection network of the co-partitioned array is omitted here for the sake of brevity (see [5]).

Example: Let us consider the partitioning of the processor space in Fig.(3). The small rectangles symbolize the LSGP-partitions. The parameters are $\mathbf{N} = (\mathbf{n}_1\ \mathbf{n}_2)$ and $\vartheta_1^{LS} = \vartheta_2^{LS} = 2$, $\boldsymbol{\Theta}^{LS} = diag(\vartheta_i^{LS})$. The large rectangles symbolize the LPGS-partitions. The parameters are $\vartheta_1^{LP} = \vartheta_2^{LP} = 2$, $\boldsymbol{\Theta}^{LP} = diag(\vartheta_i^{LP})$. The tasks of the four processing elements of the LSGP-partitions are scheduled sequentially on the black drawn processing element. The shaded processing elements are not implemented. The LPGS-partitions (i.e. large rectangles) are scheduled sequentially. Hence, the resulting array consists of two times two processing elements. •

The size L_{loc} of the local memory (i.e. the number of registers) of each processing element of the array imposed by the co-partitioning is directly proportional to the size of the LSGP-partitions. The size is given by $L_{loc} = |\mathcal{D}| \det \boldsymbol{\Theta}^{LS}$ (if $\boldsymbol{\tau}\mathbf{d}$=1). The I/O-behaviour between the array and the peripheral system has been considered in [4]. The number of variables which have to be transferred between the array and the background memory in the time interval required for the processing of a computation $\mathbf{i} \in \mathcal{I}$ is indirectly proportional to the size $\det \boldsymbol{\Theta}^{LS}$ of the LSGP-partitions. Hence, we have to search a trade-off between the implemented size L_{loc} of the local foreground memory and the access behaviour of the array and to the background memory.

3.3 Nested partitioning

As we have seen in the example given in section 2, a three-fold co-partitioning is necessary to scale the utilization of the cache and RAM level. We introduce a generalized hierarchical partitioning, called nested-partitioning, to formalize the proposed method. Each system level j is constructed of two sub-levels according to the co-partitioning policy. Hence, a mapping of an algorithm onto a system with a hierarchy of h levels requires a partitioning into $q_{max} = 2h$ sub-levels.

Definition 5 (Nested-Partitioning Scheme) *The nested-partitioning scheme is defined by the partitions $\mathcal{I}_\pi^{(q)}$ at each sublevel and the set $\mathcal{I}_\pi^{(q_{max}+1)}$. A partition at sublevel q is defined by:*

$$\mathcal{I}_\pi^{(q)} = \{\mathbf{i}^{(q)} | \mathbf{i}^{(q)} = \mathbf{i}^{(q+1)} + \kappa_n \mathbf{u} + \mathbf{G} \sum_{j=1}^{q} \left(\prod_{l=0}^{j-1} \boldsymbol{\Theta}^{(l)} \right) \boldsymbol{\kappa}^{(j)}, \mathbf{i}^{(q)} \in \mathcal{I}\}, \tag{11}$$

with $0 \le \kappa_z^{(j)} < \vartheta_z^{(j)}$, $z = 1, ..., n-1$, $\boldsymbol{\Theta}^{(j)} = diag(\vartheta_z^{(j)})$, $\boldsymbol{\Theta}^{(0)} = \mathbf{I}$. The set of (reference) points $\mathcal{I}_\pi^{(q_{max}+1)} = \{\mathbf{i}^{(q_{max}+1)}\}$ is given according to (5), where $\boldsymbol{\vartheta}$ in the vector on the right hand side in (6) has to be replaced by $\boldsymbol{\vartheta}^{(q_{max})} = \left(\prod_{l=1}^{q_{max}} \vartheta_1^{(l)} \dots \prod_{l=0}^{q_{max}} \vartheta_{n-1}^{(l)} \right)^t$.

We define that all LSGP-paramaters in (11) are represented by the odd indices q and the LPGS-parameters by the even indices q .

Now, we can turn to the consideration of the cache and RAM level utilization.

3.4 Cache utilization

As we have seen in the example given in section 2, we have to apply a two-fold co-partitioning, i.e. a nested-partitioning according to (11) with $q = 4$, for an improvement of the cache utilization. The parameter set is given by $\boldsymbol{\Theta}^{(1)} = \boldsymbol{\Theta}^{LS}, \boldsymbol{\Theta}^{(2)} = \boldsymbol{\Theta}^{LP}, \boldsymbol{\Theta}^{(3)} = diag(\vartheta_i^{(3)})$ and $\boldsymbol{\Theta}^{(4)} = \mathbf{I}$. We obtain $\vartheta_1^{(3)} = 3$ and $\vartheta_2^{(3)} = N$ for the example shown in Fig.(4). Generally, there exist $(n-1)!$ feasible loop nests (i.e. interchanges) for the sequential scheduling within the LSGP-partitions. For the example shown in Fig.(4), we have the feasible loop nests $(1, 2)$ and $(2, 1)$, where the left number represents the index of the outer loop. We chose the loop nest $(2, 1)$, i.e. the loop associated with $\mathbf{n}_1$ is the inner loop and the loop associated with $\mathbf{n}_2$ is the outer loop. With a load $L(\mathbf{n}_j) = 10^3$ for a step in direction $\mathbf{n}_j, j = 1, 2$, we obtain for the loop nest $(2, 1)$ a load $\hat{L} = L(\mathbf{n}_1) + L(\mathbf{n}_2)\vartheta_1^{(3)}$. If this load has to be exchanged only via a associated cache of the size $L_{cache} = 4 * 10^3$, then we have to ensure, that $L_{cache} \ge \hat{L}$. Hence, we achieve the condition $\vartheta_1^{(3)} \le 3$. The parameter $\vartheta_2^{(3)}$ is only bounded by the extend of the processor space [3]. For the considered example follows $\vartheta_2^{(3)} = N$.

We associate to each direction $\mathbf{n}_j$ a weighting function w_j, which depends on the selected loop nest and on the size of the LSGP-partitions of the second co-partitioning (see Fig.(4)). Hence, we obtain the condition

$$L_{cache} \ge \sum_{j=1}^{n-1} w_j L(\mathbf{n}_j) \tag{12}$$

for the selection of the parameters $\vartheta_i^{(3)}$ for a given cache size L_{cache}. The term on the right hand side of (12) represents the maximum number of variables reused during the processing of a partition in the second co-partitioning. Condition (12) ensures that the internal communication (data transfers within the partition) can be realized via the cache, i.e. without access to the RAM or mass storage.
The order of the $m = n - 1$ loops in a loop nest is defined by the sequence of numbers $o(i), i = 1, \ldots, m$, where $o(1)$ represents the outermost loop and $o(j)$ represents a more outer loop than $o(j+1)$. Furthermore holds {o(i), i=1,...,m}={1,...,m}. The weighting functions w_j are given by:

$$w_{o(i)} = \prod_{j=i+1}^{m+1} \vartheta_{o(j)}^{(3)}, \quad \vartheta_{o(m+1)}^{(3)} := 1 \tag{13}$$

Example: For a loop order $o(1) = 2, o(2) = 1$, $o(3) = 3$, the weighting functions

$$w_{o(1)=2} = \vartheta_{o(2)}^{(3)} * \vartheta_{o(3)}^{(3)}, \; w_{o(2)=1} = \vartheta_{o(3)}^{(3)}, \; w_{o(3)=3} = 1$$

result. It follows

$$\hat{L} = L(\mathbf{n}_1)\vartheta_3^{(3)} + L(\mathbf{n}_2)\vartheta_1^{(3)} * \vartheta_3^{(3)} + L(\mathbf{n}_3)$$

for the load. •

The numerical determination of the loads L has been given in [3]. Obviously, there may exist interconnections which are a linear combination of the $\mathbf{n}_j$. The derivation of the weighting functions w for such an interconnection has been given in [3].
Furthermore, we have to consider the case, that for all loads L hold $L > L_{cache}$. Then, there exist no solution of (12) for the case of a partitioning in the processor space. We can replace the condition (12) by

$$|L_{cache} - \sum_{j=1}^{n-1} w_j L(\mathbf{n}_j)| \rightarrow \min \tag{14}$$

or we include the additional partitioning of the time axis, i.e. of the $\mathbf{u}$-axis.
The objective function in the improvement of the cache utilization is the minimization of the access to the RAM and mass storage. Hence, we have to claim that the size of the LSGP-partitions in the second co-partitioning (Fig.(4)) becomes maximum, i.e. $\prod_{i=1}^{n-1} \vartheta_i^{(3)} \rightarrow \max$. That means, that we are looking for the largest for-loop programs, which satisfy condition (12) or (14). If we include the foreground memory in the consideration, then we obtain the objective function $\prod_{j=1}^{3} \prod_{i=1}^{n-1} \vartheta_i^{(j)} \rightarrow \max$.

3.5 RAM utilization

We have to apply a three-fold co-partitioning, i.e. a nested-partitioning according to (11) with $q = 6$, for an improvement of the RAM utilization. The additional parameter set is given by $\mathbf{\Theta}^{(5)} = diag(\vartheta_i^{(5)})$ and $\mathbf{\Theta}^{(6)} = \mathbf{I}$. The parameter $\vartheta_2^{(3)}$, which was only bounded by the extend of the algorithm's index space in the above discussed improvement of the cache utilization becomes now a variable. This variable reflects the increase of the data traffic between the cache and the RAM level, which occurs with the improvement of the RAM utilization. We obtain $\vartheta_2^{(3)} = N'/2$, $\vartheta_1^{(5)} = N'$ and $\vartheta_2^{(5)} = 1$ for the example shown in Fig.(5). The determination of the parameters occurs analogous to the cache level. We obtain the conditions

$$L_{RAM} \geq \sum_{j=1}^{n-1} w_j^{(R)} L^{(R)}(\mathbf{n}_j), \tag{15}$$

$$|L_{RAM} - \sum_{j=1}^{n-1} w_j^{(R)} L^{(R)}(\mathbf{n}_j)| \rightarrow \min \tag{16}$$

analogous to (12) and (14), where L_{RAM} represents the available size of the RAM. The objective function in the improvement of the RAM utilization is the minimization of the access to the mass storage. Hence, we have to claim that the size of the LSGP-partitions in the third co-partitioning (Fig.(5)) becomes maximum, i.e. $\prod_{i=1}^{n-1} \vartheta_i^{(5)} \rightarrow \max$. That means, that we are looking for the largest for-loop programs, which satisfy condition (15) or (16). If we include the foreground memory and cache in the consideration, then we obtain the objective function $\prod_{j=1}^{5} \prod_{i=1}^{n-1} \vartheta_i^{(j)} \rightarrow \max$. Since we intend to improve the utilization of the cache and RAM, we have to include here as second objective function $\prod_{i=1}^{n-1} \vartheta_i^{(3)} \rightarrow \max$. This represents the cache utilization as discussed above.

Obviously, the example given in section 2 can also be solved by a two-fold co-partitioning, where the optimization of the cache and RAM utilization has to be realized by the parameters $\vartheta_i^{(3)}$. However, the optimization of the RAM utilization can be realized without an influence on the cache utilization for algorithms with an index space of dimension $n > 3$. One parameter $\vartheta_i^{(3)}$ is bounded by the extend of the algorithms index space. Therefore, the optimization of the cache utilization causes a reduction of the dimension of the for-loop program, which represents the scheduling of the co-partitions, by one. (cf. Fig.(4b) symbolizes the one-dimensional scheduling of the two-dimensional co-partitions Fig.(4a).) We obtain a for-loop program, which represents the scheduling of the co-partitions after the second co-partitioning, of dimension $n \geq 2$ for an algorithm with index space of dimension $n > 3$. Such a program of dimension $n \geq 2$ can be partitioned for the improvement of the RAM utilization analogous to the cache problem without an influence onto the parameters $\vartheta_i^{(3)}$. Hence, the application of a six-fold partitioning for the RAM utilization establishes the more general model.

4 Concluding Remarks

The proposed memory utilization procedure can be stated as follows: The size of the implemented foreground memory (number of registers) defines the number of accesses of the array (or single processor) to the background memory in given time interval. This dependency has been discussed in [4]. The foreground memory allocation is realized by a co-partitioning. A repeated co-partitioning of the resulting program is performed to minimize the number of accesses to the RAM and mass storage level during the processing of the algorithm. This number of accesses depends directly on the size of the implemented cache memory. A third co-partitioning can be performed to minimize the number of accesses to the mass storage level during the processing of the algorithm. This number of accesses depends directly on the size of the implemented RAM memory.

The numerical values for the number of accesses to the different memory levels during the processing of the algorithm can be achieved by means of the formulas given in [4, 5] (see [3]). The computation of the I/O access pattern of the hierarchical partitioned for-loop program has been given in [2]. The program transformation of the initially given for-loop program by co-partitioning has been given in [4].

We proposed a mathemtical framework for the hierarchical partitioning of systems of URE's. However, the most algorithms represents conditional URE's. The index space of conditional URE's is the union of all conditional spaces. The parameters of the partitioning have to be adjusted to the worst case over all conditional spaces to achieve a homogeneous partitioning for the whole algorithm.

References

[1] J. Bu. *Systematic design of regular VLSI processor arrays.* PhD thesis, Delft University of Technology, 1990.

[2] U. Eckhardt. I/O management for hierarchically structured array architectures. In *Proc. 5th Euromicro Workshop on Parallel and Distributed Processing*, pages 205 – 210, 1997.

[3] U. Eckhardt. Memory management by partitioning. Technical Report SFB 358-A1-1/97, Deutsche Forschungsgemeinschaft, TU Dresden, 1997.

[4] U. Eckhardt and R. Merker. Co-partitioning- a method for hardware/software codesign for scalable systolic arrays. In R. Hartenstein and V. Prasanna, editors, *Reconfigurable Architectures*, pages 131 – 138. ITpress, 1997.

[5] U. Eckhardt and R. Merker. Scheduling in co-partitioned array architectures. In *Proc. Int. Conf. on Application-Specific Array Processors (ASAP)*, 1997.

[6] Y.-T. Hwang and Y. Hu. Novel scheduling scheme for systolic array partitioning problem. In *Proc. IEEE Workshop on VLSI Signal Processing*, pages 355 – 364, 1992.

[7] K. Jainandunsing. Optimal partitioning scheme for wavefront/systolic array processors. In *Proc. IEEE Symp. on Circuits and Systems*, 1986.

[8] R. Karp, R. Miller, and S. Winograd. The organization of computations for uniform recurrence equations. *Journal of the ACM*, 14(3):563 – 590, July 1967.

[9] D. Moldovan and J. Fortes. Partitioning and mapping of algorithms into fixed-size systolic arrays. *IEEE Trans. Computers*, C35(1):1 – 12, Jan. 1986.

[10] J. Moreno and T. Lang. Matrix computations on systolic-type meshes. *IEEE Computer*, 23(4):32 – 51, April 1990.

[11] J. Navarro, J. Llaberia, and M. Valero. Partitioning: an essential step in mapping algorithms into systolic array processors. *IEEE Computer*, 20(7):77 – 89, July 1987.

[12] H. Nelis and E. Deprettere. Automatic design and partitioning of systolic/wavefront arrays for VLSI. *Circuits systems signal processing*, 7(2):235 – 252, 1988.

[13] J. Teich. *A compiler for application specific processor arrays.* PhD thesis, Verlag Shaker, 1993.

[14] J. Teich and L. Thiele. Partitioning of processor arrays: a piecewise regular approach. *INTEGRATION, the VLSI journal*, 14(3):297 – 332, 1993.

[15] J. Teich, L. Thiele, and L. Zhang. Scheduling of partitioned regular algorithms on processor arrays with constrained resources. In *Proc. Int. Conf. on Application-Specific Array Processors (ASAP)*, 1996.

[16] L. Thiele and V. Roychowdhury. Systematic design of local processor arrays for numerical algorithms. In E. Deprettere and A.-J. van der Veen, editors, *Algorithms and parallel VLSI architectures*, volume A, pages 329–339. Elsevier Science Publisher B.V., 1991.

[17] I. Verbauwhede, F. Catthoor, J. Vandevalle, and D. H. Background memory synthesis for algebraic algorithms on multiprocessor DSP chips. In *VLSI 89*, pages 209–218. Elsevier Science Publisher B.V.

[18] I. Verbauwhede, F. Catthoor, J. Vandevalle, and D. H. In-place memory management of algebraic algorithms on application specific ics. In *Algorithms and Parallel VLSI Architectures*, pages 353–362. Elsevier Science Publisher B.V.

Architectural Exploration and Optimization of Local Memory in Embedded Systems*

Preeti Ranjan Panda Nikil D. Dutt Alexandru Nicolau

Department of Information and Computer Science
University of California, Irvine, CA 92697-3425, USA

Abstract

Embedded processor-based systems allow for the tailoring of the on-chip memory architecture based on application-specific requirements. We present an analytical strategy for exploring the on-chip memory architecture for a given application, based on a memory performance estimation scheme. The analytical technique has the important advantage of enabling a fast evaluation of candidate memory architectures in the early stages of system design. Our experiments demonstrate that our estimations closely follow the actual simulated performance, at significantly reduced run times.

1. Introduction

Increasing design complexity and shrinking product design cycle times have fueled the need for design reuse in the IC-design industry. Reuse is enabled by modern design libraries, which frequently consist of pre-designed megacells such as microprocessor cores, memories, numeric coprocessors, and modules implementing standardized functions such as JPEG. For example, the CW33000 processor core from LSI Logic, the TMS320 series of DSP processors from Texas Instruments, the StrongARM processor from Advanced RISC Machines are available in the form of *embedded cores*, in addition to the standard packaged form. Memory-based mega-modules are widely used in the industry today in the form of cache, Scratch-pad SRAM [8] and embedded DRAM [18].

An embedded processor-based system allows for customization of on-chip memory configuration according to the requirements of a specific application. This is an important feature, because a large portion of typical (packaged, off-the-shelf) microprocessors is occupied by cache memory. However, if an analysis reveals that a smaller on-chip memory results in a satisfactory performance, the resulting die area could be reduced by using the processor core with an appropriately sized on-chip memory. The large body of work on cache organization in the traditional computer architecture domain has focussed on improving the average cache performance over a large variety of applications, relying on benchmark suites such as SPEC [16]. The impact of cache memory parameters on program behavior has been widely studied in the past [14]. Techniques for modeling cache features and utilizing them in program transformations have been reported in [1, 19]. However, individual applications have widely varying memory characteristics, and in an application-specific IC, it is essential to optimize performance by tailoring the on-chip memory organization to the requirements of a given application. We address this issue in this paper.

In high-level synthesis of application-specific systems, researchers have addressed the problem of generating efficient multiport memory organizations for storing behavioral scalar and array variables [17, 15]. A strategy for alignment of behavioral arrays in off-chip memory in order to improve cache performance was reported in [10]. The problem of memory packing – determining the memory configuration for logical behavioral arrays based on a library of physical memory components has been addressed in [6, 5]. In the PHIDEO synthesis system [7] multiport memory allocation is performed for multidimensional data streams, taking interconnect costs into account. The CATHEDRAL system [3] uses a data-flow analysis technique to determine a memory architecture consisting of one or more (possibly multiport) memories satisfying a given timing constraint. However, the above works have not addressed the memory architecture issues involved when the data size is too large to be stored on-chip.

We present an exploration strategy for determining an efficient on-chip memory architecture – characterized by Scratch-pad memory size and cache parameters – based on an analysis of a given application, so as to reduce off-chip

*This work was partially supported by grants from ARPA (MDA904-96-C-1472), NSF(CDA-9422095), and ONR(N00014-93-1-1348).

memory traffic. The advantage of such an analytical approach over a simulation-based strategy is that, it is possible to do a fast comparison of the expected performance resulting from many different memory architectures, whereas simulation of the application for all the possible memory configurations could be prohibitively expensive, and impractical for purposes of exploration.

2. Architecture

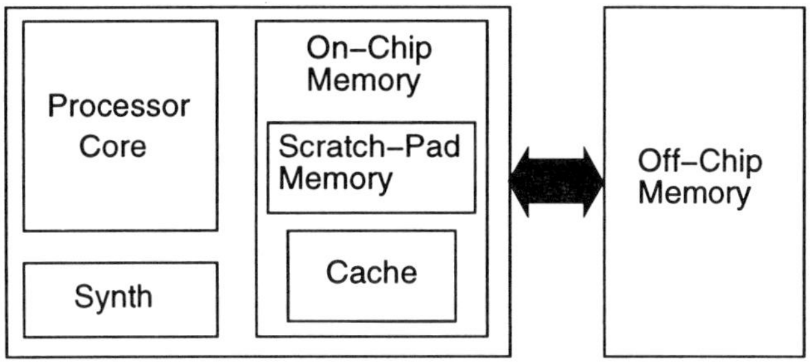

Figure 1. Embedded Processor-based Architecture

Figure 1 shows a simplified view of an embedded processor core-based system, consisting of a processor core, on-chip memory, and synthesized block (*Synth*), interfacing with off-chip DRAM. The synthesized hardware block *Synth*, often the result of behavioral synthesis, performs the functions specific to the application that are mapped to hardware, possibly for performance considerations. This decision of mapping different parts of a design into hardware and software is taken in a prior Hardware/Software partitioning step. In this paper, we assume that the partitioning has already been performed. The on-chip memory can be implemented as a combination of cache and *Scratch-pad SRAM*. In the current work, we limit our attention to on-chip data memory.

Data cache is fast, on-chip memory forming an interface between the processor and the off-chip DRAM, that reduces the effective memory access time by storing recently accessed data [14]. Internally, the cache is divided into blocks, or *cache lines*, which constitute the smallest unit of interaction between the cache and the off-chip memory.

Scratch-pad SRAM is on-chip memory, to which the assignment of data is compiler-controlled. The overall architecture is described in detail in [13]. In brief, a portion of the total data memory space is mapped to on-chip SRAM (typically used to store critical data), with the advantage of guaranteed fast access, unlike the cache, where hardware-controlled storage and replacement strategies could flush out data into the off-chip memory, resulting in cache misses that stall the processor. In some modern embedded systems such as the F/X256 graphics controller from Silicon Magic Corporation, the Scratch-pad memory assumes the form of *embedded DRAM* [18].

3. Illustrative Example

We illustrate our strategy for architectural exploration and optimization of on-chip memory on CONV, a convolution program frequently used in Image Processing tasks such as edge detection, regularization, and morphological operations [2]. The code for CONV is shown below:

```
N = 128; M = 4; NORM = 16;
int source[N][N], dest [N][N], mask [M][M];
Procedure CONV
int acc, i, j, x, y;
for (x = 0; x < N – M; x++)
    for (y = 0; y < N – M; y++) {
        acc = 0;
        for (i = 0; i < M; i++)
            for (j = 0; j < M; j++)
                acc = acc + source[x+i][y+j] * mask[i][j];
        dest[x+M/2][y+M/2] = acc/NORM;
    }
```

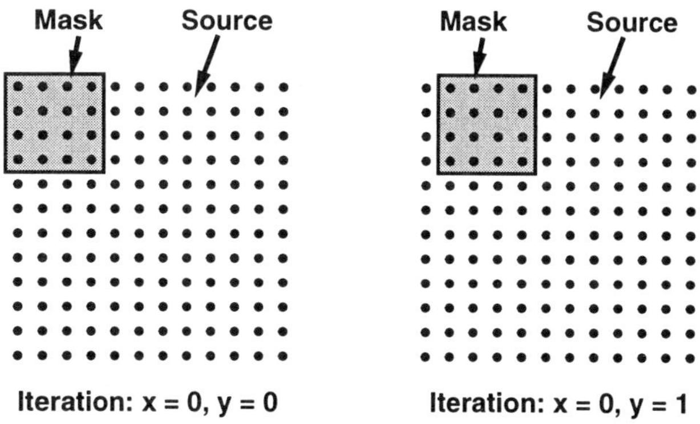

Figure 2. Memory access pattern in CONV example

A small (4×4) matrix of coefficients, *mask*, slides over the input image, *source*, covering a different 4×4 region in each iteration of y, as shown in Figure 2. In each iteration, the coefficients of *mask* are combined with the region of the image currently covered, to obtain a weighted average, and the result, *acc*, is assigned to the pixel of the output array, *dest*, in the center of the covered region.

We first attempt to solve the following problem: given a maximum amount, say 4 KB of on-chip memory space, find an efficient utilization of the space, in terms of data cache size, cache line size, and Scratch-pad memory size, with the objective of minimizing the number of processor cycles required to access the arrays from memory. We assume that scalar variables are assigned to registers; the data cache is direct-mapped and write-through [14]; and the largest allowed cache line size is 128 Bytes.

We note that, since we use a write-through cache, in which memory writes do not interfere with the cache contents in case of cache misses, the access to *dest* does not cause cache conflicts. However, if the two arrays *source* and *mask* were to be accessed through the data cache, the performance would be affected by cache conflicts. Further, an associative cache, by itself, will not eliminate the problem in general, because most practical caches have a limited associativity, and the general situation might require an associativity as large as the number of conflicting arrays.

The conflict problem can be solved by storing the small *mask* array in the Scratch-pad memory. This assignment eliminates all conflicts in the data cache – the data cache is now used for memory accesses to *source*, which are very regular. Also, since $M(= 4)$ rows of the *source* array are *active* at any point in time, the data cache need be only as large as 4 rows of the source array. [1] Thus, we select a data cache of size $M \times N = 4 \times 128 = 512$ words = 2 KB. Since the accesses to *source* have good spatial locality, we select the largest allowed cache line, i.e., 128 bytes. The Scratch-pad memory size is 4×4 words = 64 Bytes.

4. Memory Architecture Exploration

In our formulation, a local (on-chip) memory architecture for an application is defined as a combination of:

- The total size of on-chip memory used for data storage.
- The partitioning of this on-chip memory into: (1) Scratch-pad SRAM, characterized by its size; and (2) data cache, characterized by the cache size; and the cache line size.

The basic algorithm for memory architecture exploration is summarized below:

Algorithm *MemExplore*
L_1: **for** on-chip memory size T (in powers of 2)
  L_2: **for** cache size C (in powers of 2, $< T$)
    SRAM Size $S = T - C$
    DataPartition(S)
    L_3: **for** line size L (in powers of 2, $< C$, $<$ *MaxLine*)
      Estimate Memory Performance
  Select (C, L) that maximizes performance

For each candidate on-chip memory size T (loop L_1), we consider different divisions of T (loop L_2) into cache (size C) and Scratch-pad SRAM (size $S = T - C$), selecting only powers of 2 for C. Procedure *DataPartition* is based on a technique for partitioning program variables into Scratch-pad memory and cache [13]. Scalar and array data identified to be the most critical, are assigned to the SRAM, based on the data size, the memory access frequency and the possibility of cache conflicts. In the rest of this section, we describe the memory performance estimation step. For each T, we select the (C, L) pair that is estimated to maximize performance. Finally, we perform the memory address assignment of the variables using the cache and SRAM parameters selected using the algorithms in [12] and [13].

[1] As mentioned before, the *dest* array is not allocated any storage in a write-through cache. If a *write-back* cache were used instead, we would allocate an equal cache space to *dest*, along with an appropriate memory assignment so that *source* and *dest* do not conflict in the cache [12].

4.1. Memory Performance Estimation

There is a trade-off in sizing the cache line. If the memory accesses are very regular and consecutive, i.e, exhibit spatial locality, a longer cache line is desirable, since it minimizes the number of off-chip accesses and exploits the locality by pre-fetching elements that will be needed in the immediate future. On the other hand, if the memory accesses are irregular, or have large strides, a shorter cache line is desirable, as this reduces off-chip memory traffic by not bringing unnecessary data into the cache. The maximum size of a cache line is the DRAM page size, which is usually less than 1 KB = 256 $(= 2^8)$ words for most modern DRAMs. Thus, there are typically a maximum of 9 alternatives for the cache line size (which is usually a power of 2).

Suppose there are N scalar variables stored in off-chip memory, accessed M times in the program. We store all scalar variables in consecutive locations in memory. Since accesses to scalars invariably constitute a small fraction of the total accesses, we make the simplifying assumption that there is only one cache miss (a *compulsory* miss that occurs when the variable is first accessed into the cache) for every cache line containing scalars. Although this assumption looks too optimistic, it is actually reasonable when combined with our partitioning strategy for scalar and array variables – most of the scalars get mapped to the register file and Scratch-pad memory, and not to the cache. Since there are N scalars, we require $\lceil \frac{N}{L} \rceil$ cache lines for them for a cache line size of L, i.e., there are $\lceil \frac{N}{L} \rceil$ cache misses, and consequently, $M - \lceil \frac{N}{L} \rceil$ cache hits for the M accesses. A memory access resulting in a cache hit requires one cycle, while a cache miss entails a delay of $(K + L)$ processor cycles, where K is a constant (usually 10–20 in modern processors [14]). Thus, the total number of processor cycles required for the M accesses to scalar variables is:

$$\text{Cycles (Scalars)} = (K + L)\left\lceil \frac{N}{L} \right\rceil + M - \left\lceil \frac{N}{L} \right\rceil \quad (1)$$

We determine an estimate of the processor cycles required to access the array elements by first dividing the application program into loop nests. Straight line code is

```
for i = 1 to M − 1 step 1
   for j = 1 to M − 1 step 1
      A[i][j] = A[i][j] + A[i − 1][j] + A[i + 1][j]+
         A[i][j − 1] + A[i][j + 1] + B[i] + C[j][i]
```

(a)

```
L1 : for i1 = l1 to h1 step s1
   L2 : for i2 = l2 to h2 step s2
      ...
      Lm : for im = lm to hm step sm
         Read a[i1][i2]...[im]
         ...
         Ln : for in = ln to hn step sn
            Read b[i1 + k1][i2 + k2]...[in + kn]
            Read b[i1 + k'1][i2 + k'2]...[in + k'n]
```

(b)

Figure 3. (a) Example loop (b) General n-level loop nest

treated as a singly-nested loop with an iteration count of one. Multi-dimensional arrays are assumed to be stored in row-major format.

Consider the example loop shown in Figure 3(a). $B[i]$ is reused in different j-iterations, so it is moved up into the i-loop. $A[i][j], A[i][j-1]$, and $A[i][j+1]$ exhibit *group-spatial* reuse [19] – the cache line accessed by one reference will usually also contain the data for the others. $A[i-1][j]$ and $A[i+1][j]$ have *self-spatial* reuse [19] – each can reuse the cache line it accessed in the previous j-iteration. Finally, reference $C[j][i]$ has no reuse. The memory references are grouped into *reuse equivalence classes* – each class consists of memory references that exhibit self-spatial or group-spatial reuse. This classification is used in [19] to aid in loop transformation procedures, such as skewing, reordering, etc., by enabling a comparison of the reuse properties of different candidate transformations. We propose a refinement of the above procedure in order to utilize the reuse analysis for our line size selection problem.

4.1.1 Refinement to self-spatial locality

In Figure 3(a) reference $B[i]$, after being moved to the i-loop, was assumed to have spatial reuse. However, this reuse can occur only if the cache line corresponding to $B[i]$ is still present in the cache when the next i-iteration begins, and has not been flushed out by the intervening accesses.

We generalize the above reuse condition for the example n-level nested loop of Figure 3(b). We assume that the loop bounds, and branch probabilities for conditionals are statically known, as is the usual case in many embedded applications. We use the following locality criterion: spatial locality for the reference $a[i_1][i_2]\ldots[i_m]$ in the level-m loop (L_m) can be exploited if the total number of memory accesses in inner loops (i.e., loops $L_{m+1}\ldots L_n$) is less than the cache size. Let the number of memory references at loop level j be c_j. For example, in Figure 3(b), $c_m = 1$ (for the single memory read), and $c_n = 2$ (for the two memory reads). The number of iterations (r_j) of loop level L_j is given by: $r_j = \left\lceil \frac{h_j - l_j + 1}{s_j} \right\rceil$. Thus, the sufficient condition for utilizing locality for the reference at nesting level m is:

$$\sum_{i=m+1}^{n} c_i \left(\prod_{j=m+1}^{i} r_j \right) \leq \mathit{CacheSize} \tag{2}$$

Note that the condition above, which involves the *number* of elements accessed, is an approximation of the cache behavior, for it assumes a fully associative cache with a perfect replacement policy. We incorporate the effect of cache conflicts that occur in limited associativity caches in Section 4.1.3.

4.1.2 Refinement to group-spatial locality

In Figure 3(a), reference $A[i-1][j]$ and $A[i][j]$ are assumed to have no spatial or temporal locality because they are in different rows. However, this is a pessimistic assumption. If one complete row of A fits in the cache, then the data read by $A[i][j]$ in one iteration can be reused by the $A[i-1][j]$ reference in the next i-iteration.

We formalize the above observation into a general condition for predicting reuse for the two b-references in Figure 3(b). We first determine a feasibility condition that needs to be met if the two references are to access the same cache line in different iterations. The sharing can occur only if all the index expressions in the higher dimensions $(1\ldots n-1)$ match exactly for different values of the loop index, and the reuse-dimension (lowest dimension) resolves to expressions that differ in less than the cache line size. For example, we require that $(i_1 + k_1)$ in one iteration, say $i_1 = I$, be equal to $(i_1 + k'_1)$ in some other iteration, say $i_1 = I + s_1 \cdot t$, for an integer t, since the two iteration numbers are separated by a multiple of the loop stride, s_1. We need to have: $I + k_1 + s_1 \cdot t = I + k'_1$, i.e., $k'_1 - k_1 = s_1 \cdot t$, i.e., $(k_1 - k'_1) \bmod s_1 = 0$.

Generalizing for dimensions $[1\ldots n-1]$, we have:

$$\forall j \in [1\ldots n-1], (k_j - k'_j) \bmod s_j = 0 \tag{3}$$

For the reuse dimension (lowest dimension, indexed by i_n), we do not need an exact match, but only need that the expressions differ in less than the cache line size L. Thus, for two iterations: $i_n = I$ and $i_n = I + s_n \cdot t$, we need to have:

$$(I + k_n + s_n \cdot t) - (I + k'_n) \leq L \tag{4}$$

i.e.,

$$k_n - k'_n - L \leq s_n \cdot t \quad (5)$$

To be applied as a feasibility test, this can be rephrased as:

$$\exists l \in [1 \ldots L], \text{ such that } (k_n - k'_n - l) \bmod s_n = 0 \quad (6)$$

Further, we need to ensure that the number of elements brought into the cache between the two accesses to b in Figure 3(b), is less than the cache size. For the two index expressions $[i_1 + k_1]$ and $[i_1 + k'_1]$, the number of iterations of loop L_1 that elapse between the two expressions resolving to the same value is: $\left\lfloor \frac{|k_1 - k'_1|}{s_1} \right\rfloor$. Since c_1 elements are accessed at the first loop level, the number of elements from loop L_1 brought into the cache in the $\left\lfloor \frac{|k_1 - k'_1|}{s_1} \right\rfloor$ iterations is: $c_1 \cdot \left\lfloor \frac{|k_1 - k'_1|}{s_1} \right\rfloor$. The number of elements in inner loops accessed in each iteration of loop L_1 is: $\sum_{i=2}^{n} c_i (\prod_{j=2}^{i} r_j)$ (LHS of Equation (2), with $m = 1$). Thus, the total number of elements (f_1) brought into the cache before $[i_1 + k_1]$ and $[i_1 + k'_1]$ resolve to the same value, is given by:

$$f_1 = c_1 \cdot \left\lfloor \frac{|k_1 - k'_1|}{s_1} \right\rfloor \cdot \sum_{i=2}^{n} c_i \left(\prod_{j=2}^{i} r_j \right) \quad (7)$$

Summing over all the n dimensions, we have the sufficient condition to enable group-spatial reuse as:

$$\sum_{t=1}^{n} f_t \leq \mathit{CacheSize} \quad (8)$$

i.e.,

$$\sum_{t=1}^{n} \left(c_t \cdot \left\lfloor \frac{|k_t - k'_t|}{s_t} \right\rfloor \cdot \sum_{i=t+1}^{n} c_i \left(\prod_{j=t+1}^{i} r_j \right) \right) \leq \mathit{CacheSize} \quad (9)$$

We conclude that the two b-references in Figure 3(b) exhibit spatial locality (i.e., fall into the same reuse equivalence class) if they satisfy the conditions 3, 6 and 9. The equations can be generalized to the case where the array index expressions are, of a more general form: $[a_j i_j + k_j]$ (where a_j and k_j are constants), as is the case with array references in many multimedia applications. The generalization is described in [11].

4.1.3 Refinement incorporating cache conflicts

Note that Equations 2 and 9 are approximations for cache reuse, for they ignore the possibility of cache conflicts. We apply the technique in [12] to determine an estimate of the number of cache conflicts between memory references in different equivalence classes. Each conflict represents one off-chip memory access. We define *EstConflict*() to be a procedure that returns the number of processor cycles wasted due to conflicts. This is given by: *EstConflict*(j) = (# Conflicts predicted for accesses in loop L_j) $\times (K + L)$, where K and L are as defined in Equation (1).

4.1.4 Computation of Processor Cycles

Based on the analysis in the previous sections, we determine the number of processor cycles, T_{ji}, due to each reuse equivalence class R_{ji} in loop level L_j, as in the loop nest of Figure 3(b). Let the number of memory references in R_{ji} be $|R_{ji}|$. For the r_j iterations of loop L_j, the total number of memory accesses for class R_{ji} is: $r_j \cdot |R_{ji}|$, of which $\frac{r_j \cdot |R_{ji}|}{L}$ are cache misses, and the rest $(r_j \cdot |R_{ji}| - \frac{r_j \cdot |R_{ji}|}{L})$ are hits. Since a cache hit costs 1 cycle and a miss $(K + L)$ cycles, we have:

$$\begin{aligned} T_{ji} &= \frac{r_j \cdot |R_{ji}|}{L}(K + L) + r_j \cdot |R_{ji}| - \frac{r_j \cdot |R_{ji}|}{L} \\ &= r_j \cdot |R_{ji}| \cdot \left(\frac{K - 1}{L} + 2 \right) \quad (10) \end{aligned}$$

For the set E of all the equivalence classes at level j, we have, the total cycles, T_j given by: $T_j = \sum_i T_{ij}$.

The number of accesses with no reuse at loop level L_j is given by: $c_j - \sum_i |R_{ji}|$. The number of cycles, Q_j, spent in accessing elements with no reuse is given by:

$$Q_j = (K + L) \cdot \left(c_j - \sum_i |R_{ji}| \right) \quad (11)$$

Finally, the number of processor cycles, P_j, wasted due to cache conflicts for accesses in loop L_j is given by:

$$P_j = r_j \cdot \mathit{EstConflict}(j) \quad (12)$$

Thus, the total time, spent accessing memory data at level L_j is given by: $(T_j + Q_j + P_j)$. For the entire loop nest, we multiply the total cycles at each loop level by the number of times the loop L_j is executed. Thus, the total cycles, C for the loop nest under consideration is given by:

$$C = \sum_{j=1}^{n} \left((T_j + Q_j + P_j) \cdot \prod_{i=1}^{j-1} r_j \right) \quad (13)$$

To determine the number of cycles required for all memory accesses in the program, we take the cumulative estimate over all loop nests. We use this estimate in the memory performance estimation step in algorithm *MemExplore*.

5. Experiments and Results

We present some exploration results on sample application routines from the image processing and digital signal processing domain. The benchmark examples are: *Histogram* – a histogram evaluation routine, commonly used in image enhancement algorithms [4]; *Lowpass* – another image processing algorithm used for accentuating low frequencies in an image, so that the resulting image has lower changes between neighboring pixel values [9]; and *Beamformer* – a radar application, involving the summation of digitized signals from an antenna array [9]. We present below a comparison of the estimated memory cycles for the benchmarks with the actual memory cycles. The estimated cycles are determined by the computations in Section 4. The actual cycles are measured by a memory simulator we developed, which takes as input a stream of memory addresses generated during execution of a benchmark, and reports the number of accesses to off-chip memory by simulating the data cache and Scratch-pad memory. For our experiments, we used a penalty of 10 processor cycles for off-chip memory accesses (i.e., $K = 10$ in Equations 1 and 11).

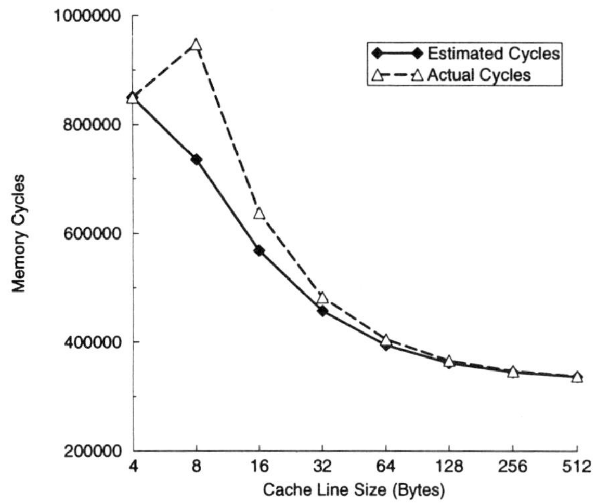

Figure 4. Variation of memory performance with line size for fixed cache size of 1 KB (*Histogram* Benchmark)

Our first experiment focusses on loop L_3 of the exploration algorithm *MemExplore*, where we study variation of the memory performance with the cache line size, for a given cache size. Figure 4 shows a comparison between the actual simulation result and the estimated number of processor cycles for the memory accesses in the *Histogram* routine, for a fixed cache size of 1 KB. We note that the estimated performance very closely follows the actual performance, except for one line size (8 bytes). The best cache line size selected by our algorithm (= 512 bytes), is also the best line size parameter determined from the simulation.

Figure 5 illustrates another slice of the exploration space on the same benchmark, where the estimated memory per-

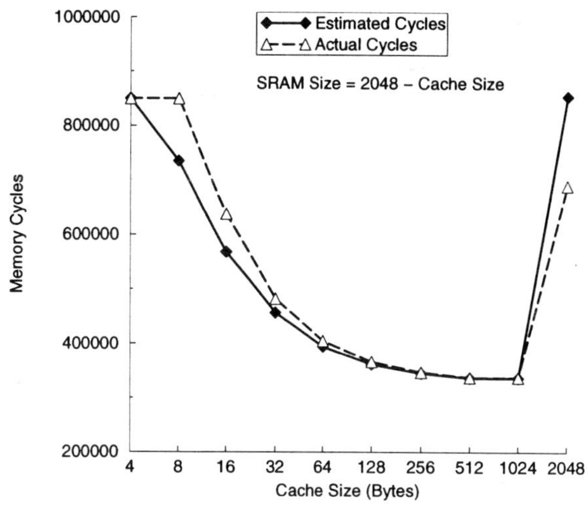

Figure 5. Variation of memory performance with different mixes of cache and Scratch-pad memory, for total on-chip memory of 2 KB (*Histogram* Benchmark)

formance for different divisions of a fixed total on-chip memory space of 2 KB into data cache and Scratch-pad memory, is compared against the simulated performance. The memory cycles plotted correspond to one iteration of the outer L_2 loop of *MemExplore*, where the best cache line size (from loop L_3) is selected for each candidate cache size. For each selected cache size, the corresponding Scratch-pad SRAM size is given by: 2 KB – cache size. The points on the left and right extremes represent divisions incurring severe cache conflicts (for cache size = 2048 Bytes, the SRAM size is 0, causing unavoidable conflicts in the cache). The estimation process gives the best division of the 2 KB space as: 1 KB cache + 1 KB Scratch-pad memory – this selection is validated with the actual simulation results, as shown in Figure 5.

Figure 6 shows the variation of the memory performance with the total on-chip memory space for the *Histogram* example. The y-axis shows the best performance obtained by any architecture (i.e., division into Scratch-pad memory and cache, as well as selection of cache line size) for a given total on-chip memory space, i.e., one iteration of the outer loop L_1 in *MemExplore*. For a given application, the variation of the memory performance with the total on-chip memory is generated as feedback to the designer. The designer can then select an appropriate total on-chip memory size, based on the value beyond which no significant improvement is predicted. In the example of Figure 6, the total size of 2 KB is a good selection, as we observe very little improvement in cycle time beyond this cache size. Figures 7(a) and (b) show the results of the exploration for the *Lowpass* and

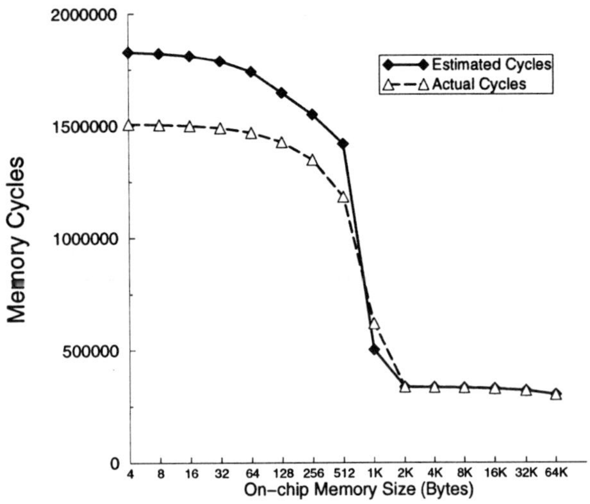

Figure 6. Variation of memory performance with total on-chip memory space (*Histogram* Benchmark)

Beamformer examples. We notice that the estimated performance curve follows the actual performance very closely. This curve can help the designer select the optimal on-chip memory size. Once the designer chooses an appropriate total memory size based on the estimation curve, the best memory architecture (consisting of the division of this space into Scratch-pad memory and cache, as well as cache line size) is automatically chosen. Note that the each point in the exploration spaces of Figures 6 and 7 corresponds to the best architecture found for the given total on-chip memory space.

The most important advantage of our analytical technique for exploring the memory performance of embedded applications is that candidate architectures can be rapidly evaluated for their memory performance. The estimation-based exploration for our experiments above required only a few seconds, which was about 1000 times faster than the simulation of the memory performance for the same set of architectures explored. This estimation capability is very important in the initial stages of system design, where the number of possible architectures is too many, and a simulation of each architecture is prohibitively expensive.

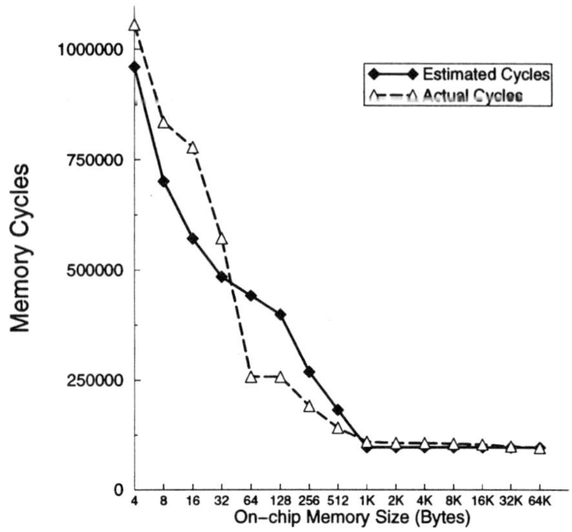

(a)

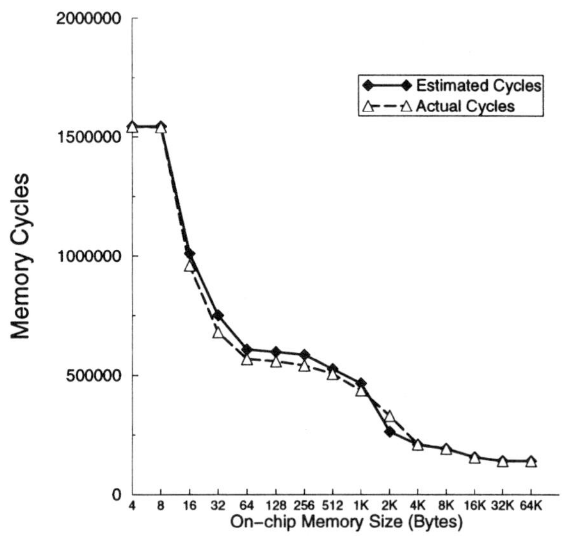

(b)

Figure 7. Memory exploration for (a) *Lowpass* and (b) *Beamformer* benchmarks

6. Conclusions

The management of local memory space is an important problem in embedded systems involving large behavioral arrays, all of which cannot be stored on-chip. The ideal local memory architecture, consisting of a judicious division of the memory into Scratch-pad SRAM (software-controlled) and data cache (hardware-controlled), as well as the cache line size, depends on the characteristics of the specific application.

We presented a strategy for exploration of on-chip memory architecture for embedded applications, based on an estimation of the memory performance. Our experiments on benchmark routines show that the estimated performance matches the actual simulated performance very closely, and is three orders of magnitude faster. Thus, the exploration technique is very useful during the early stages of system design, when a large number of different possible memory architectures need to be evaluated for their performance.

The future work for this research includes the incorporation of other cache features, such as write policy, and the incorporation of a more accurate cost function for on-chip memory area that accounts for the memory decoder area.

References

[1] A. Agarwal, M. Horowitz, and J. Hennessy. An analytical cache model. *ACM Transactions on Computer Systems*, 7(2):184–215, May 1989.

[2] P. Baglietto, M. Maresca, M. Migliardi, and N. Zingirian. Image processing on high performance RISC systems. Technical Report TR SM-IMP/DIST/08, University of Genoa, December 1995.

[3] F. Balasa, F. Catthoor, and H. D. Man. Dataflow-driven memory allocation for multi-dimensional signal processing systems. In *Proceedings of the IEEE/ACM International Conference on Computer Aided Design*, November 1994.

[4] R. C. Gonzalez and P. Wintz. *Digital Image Processing*. Addison-Wesley, 1987.

[5] P. K. Jha and N. Dutt. Library mapping for memories. In *European Design and Test Conference*, pages 288–292, March 1997.

[6] D. Karchmer and J. Rose. Definition and solution of the memory packing problem for field-programmable systems. In *Proceedings of the IEEE/ACM International Conference on Computer Aided Design*, pages 20–26, November 1994.

[7] P. E. R. Lippens, J. L. van Meerbergen, W. F. J. Verbaeghand, and A. van der Werf. Allocation of multiport memories for hierarchical data streams. In *Proceedings of the IEEE International Conference on Computer Aided Design*, November 1993.

[8] LSI Logic Corporation, Milpitas. *CW33000 MIPS Embedded Processor User's Manual*, 1992.

[9] P. R. Panda and N. D. Dutt. 1995 High level synthesis design repository. In *International Symposium on System Synthesis*, September 1995.

[10] P. R. Panda, N. D. Dutt, and A. Nicolau. Memory organization for improved data cache performance in embedded processors. In *International Symposium on System Synthesis*, pages 90–95, November 1996.

[11] P. R. Panda, N. D. Dutt, and A. Nicolau. Architectural exploration and optimization of local memory in embedded systems. Technical Report ICS-TR-97-31, University of California, Irvine, June 1997.

[12] P. R. Panda, N. D. Dutt, and A. Nicolau. Data cache sizing for embedded processor applications. Technical Report ICS-TR-97-30, University of California, Irvine, June 1997.

[13] P. R. Panda, N. D. Dutt, and A. Nicolau. Efficient utilization of Scratch-pad memory in embedded processor applications. In *European Design and Test Conference*, March 1997.

[14] D. A. Patterson and J. L. Hennessy. *Computer Organization & Design – The Hardware/Software Interface*. Morgan Kaufman, 1994.

[15] L. Ramachandran, D. Gajski, and V. Chaiyakul. An algorithm for array variable clustering. In *European Design and Test Conference*, February 1994.

[16] Standard Performance Evaluation Corporation, Fairfax. *SPEC Newsletter*, December 1991.

[17] L. Stok and J. A. G. Jess. Foreground memory management in data path synthesis. *International Journal of Circuit Theory and Applications*, 20(3):235–255, 1992.

[18] R. Wilson. Graphics IC vendors take a shot at embedded DRAM. *Electronic Engineering Times*, (938):41,57, January 27 1997.

[19] M. E. Wolf and M. Lam. A data locality optimizing algorithm. In *Proceedings of the SIGPLAN'91 Conference on Programming Language Design and Implementation*, pages 30–44, June 1991.

Session 5

System-Level Synthesis and Design

Optimization Method for Broadband Modem FIR Filter Design using Common Subexpression Elimination

Robert Pasko [1], Patrick Schaumont [2], Veerle Derudder [2], Daniela Durackova [1]
[1] Faculty of Electrical Engineering & Information Technology
Dept. of Microelectronics, Bratislava, Slovakia
[2] IMEC Leuven, Belgium

Abstract - **An approach for broadband modem FIR filter design optimization is presented. It addresses the minimization of the number of adder-subtractors used in the hardware implementation of a FIR filter (Multiple Constant Multiplication problem). The method is based on identification and elimination of n-bit pattern common subexpressions in a set of filter coefficients by means of an exhaustive search. We give an algorithm description of our solution and demonstrate the performance on selected examples. A comparison of the results obtained by other authors is made and finally optimization and synthesis results on a realistic example, a 64-tap root-raised-cosine filter with 10 bit CSD coefficients is given.**

I. Introduction

The advent of broadband modems in consumer applications such as access network communications has attracted a lot of research towards the design of high speed FIR filters. The design requirements for these filters are such that they must process wideband signals with considerable accuracy. Because of the speed requirements, programmable solutions such as DSP cores are of no use in dealing with these problems. Rather, an application specific approach in hardware is needed. Since the existing standardization efforts [1,2] specify only the shape of the filter responses, and make no statement regarding their implementation, efficient VLSI synthesis methods are needed.

In this paper, an algorithm for optimization of a transpose form FIR filter with constant coefficients by an exhaustive search is presented. Today, exhaustive search (or brute force) algorithms are often combined, or even completely replaced, by sophisticated heuristic methods to gain the performance required in complex tasks, e.g. chess game algorithms. On the other hand, assumed appropriate algorithms are used, exhaustive search can be used to solve some less complex problems with very satisfactory results. We have used an exhaustive search method to find multiple common bit patterns in a FIR filter coefficient array. By removing multiple occurrences of the same pattern, and calculating these only once, hardware is saved. For larger filters, this leads to important area reductions, as will be demonstrated. During the optimization process, several techniques are used to obtain a cheaper implementation. First, all multiplications in a FIR filter are splitted into sequences of add-shift operations using a radix-2 signed digit representation of the coefficients. A Canonic Signed Digit (CSD) representation [4] is used. It is a representation with a minimum number of non-zero bits, and therefore a minimum number of adder-subtractors are necessary for the expansion.

In the next step, the set of coefficients to implement is subject to Common Subexpression Elimination (CSE). This involves finding multiple bit-patterns in the set of coefficients and coding these repetitions as common subexpressions into the filter structure. Hardware then is saved by the reduction of the number of adders necessary. An example of CSE optimization is demonstrated in Figure 1. Common subexpression elimination technique already has been examined in various papers [5-8]. We briefly indicate the approaches that are followed. In [5], iterative elimination of only 2-bit common subexpressions is proposed, in [6] an iterative matching algorithm is used, in [7] a matrix-based approach for subexpression sharing is presented and in [8] a reduced adder graph algorithm is described. Some of these algorithms use heuristics to derive a cheaper implementation. In our method, we use an exhaustive search over the coefficient set to find the common subexpressions in filter coefficients. Thanks to an efficient coding of the search algorithm, a better optimization is obtained in combination with efficient runtimes.

1080-1820/97 $10.00 © 1997 IEEE

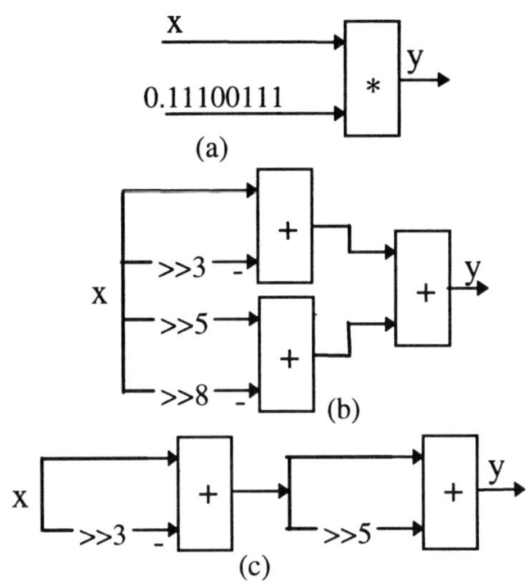

(a) multiplication with coeff. 0.11100111

(b) CSD add-shift expansion (1.00$\underline{1}$01001$\underline{}$)

(c) add-shift after 100$\underline{1}$ pattern elimination

Figure 1. CSE optimization example

II. Method Overview

The basic steps of our method are shown in Figure 2. As input, a set of filter coefficients in floating point representation is used. During the first step, each of these coefficients is converted to its closest CSD representation according to user-selected parameters. This step is examined in section III.

Next, the CSD coefficients set is sent to CSE. During this step, multiple bit-patterns are eliminated from the set. Based on the reduced specification, VHDL behavioral code of the FIR filter is generated. CSE is discussed in detail in section IV. The performance of our algorithm is demonstrated with two examples in section V.1. In the first one, different coefficient sets are optimized and ratios between the number of adders-subtractors in initial and optimized sets are calculated. In the second example, a Root-Raised-Cosine 64-tap FIR filter with 10bit CSD coefficients is synthesized. Section V.2 then compares the obtained results to that of other authors. Finally, conclusions are drawn in section VI.

III. Quantization and CSD Coding

The initial specification for the transposed-form FIR filter is given as a set of floating point coefficients, each one representing a tap coefficient. The first task towards implementation is to quantize these into finite wordlength CSD representation. The number of bits required depends on the application at hand. For the case of modem design, some design rules can be found in [4]. A canonic signed digit representation is defined as a representation with a minimal number of non-zero digits for which no two non-zero digits are adjacent. The efficiency of CSD representation for constant multiplication is demonstrated in Figure 1(b), where only 3 adders instead of 5 are necessary. A floating point coefficient is converted during quantization to its closest CSD representation according to three criterions : the coefficient wordlength, the maximum number of non-zero bits, and the allowed quantization error. No other optimization is performed during this step. Several possible optimization methods are discussed in [4].

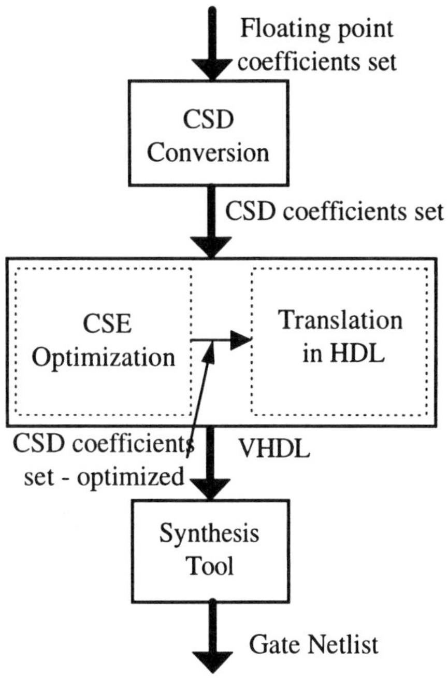

Figure 2. Method flow-graph

IV. Common Subexpression Elimination in FIR Filters

During CSE optimization, redundant hardware is minimized by finding repeating structures and implementing them only once. A demonstration example of CSE is shown in Figure 3a 3-tap transpose-form FIR filter is depicted in Figure 3a, with coefficients in Tab 1.

h_0	$0.100\underline{1}0101_B$
h_1	$1.00100\underline{1}00_B$
h_2	$1.0\underline{1}0\underline{1}0000_B$

Table 1. Filter Coefficients in filter 1.

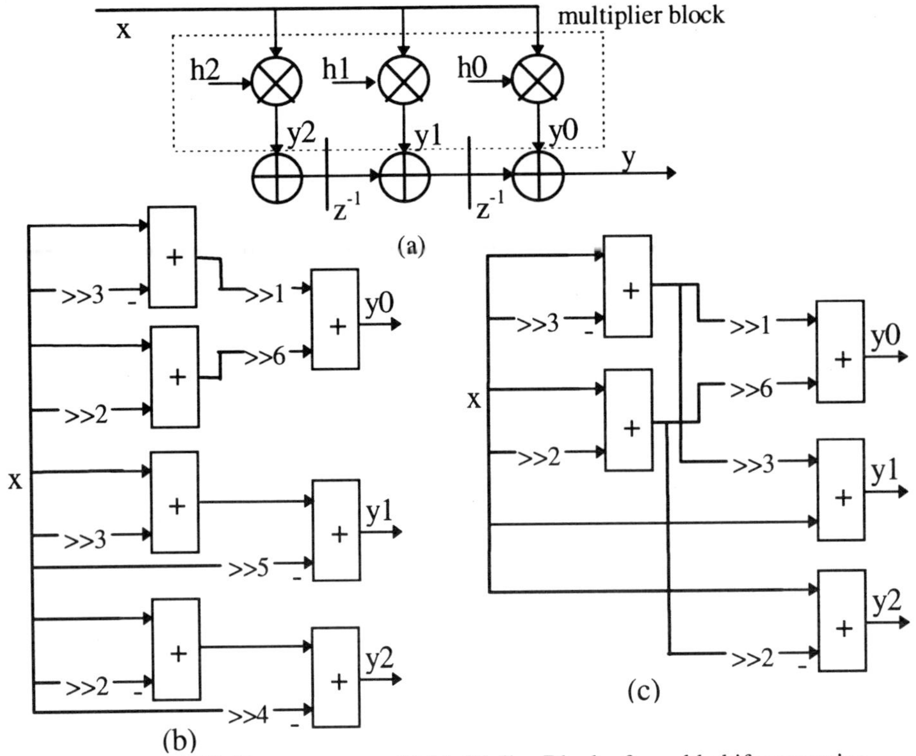

(a) Transpose FIR filter structure (b) Multiplier Block after add-shift expansion
(c) Hardware reduction after CSE optimization

Figure 3. CSE example

Figure 3(b) shows the implementation after CSD add shift expansion. In this implementation, the patterns 100$\underline{1}$ and 101 both occur twice, and this is detected by our algorithm. The optimized structure is shown in Figure 3(c). We indicate that the relative position of a bit pattern is of low importance since all shifts are considered hardwired. During CSE, an exhaustive search for common n-bit patterns is performed. To do this, the frequency of each subexpression pattern is determined, and next the most common pattern is chosen and eliminated. This process is repeated until there does no longer exist a pattern with frequency higher than 1. The complete block-diagram of our algorithm is presented in Figure 4. A parameter n is required, which must be in the range <2, (N+1)/2>. The upper bound is equal to the maximum number of non-zero bits in an N-bit CSD number.

During statistics creation, all possible n-bit patterns of each coefficient are identified and included into the frequency statistics. Next, the set of frequency statistics is searched for the most common pattern and, assumed a frequency is higher than one, all occurrences of the selected pattern are eliminated. Afterwards, the selected pattern is added as a new coefficient at the end of the coefficient set. Also, because elimination of one pattern must influence the frequency statistics of the others, the statistics must be re-evaluated. This procedure is repeated for *n, n-1, n-2 ... 2* bit patterns.

We will now discuss the important steps of the optimization procedure in more detail. These include:

1. Pattern identification

2. Pattern selection

3. Searching algorithm

We will also indicate the efficiency of our algorithm.

IV.1 Pattern Identification

In an exhaustive search, all possible combinations of n-non-zero bit patterns in a coefficient must be examined. Since a pattern can only be eliminated once, we must also detect the occurrence of the same patterns within each other. For example, the valid 2-non-zero bit patterns of coefficient 010$\underline{1}$0$\underline{1}$0$\underline{1}$ are in the Table 2. The frequency of coefficient $\underline{1}$0$\underline{1}$ is one, because the second instance contains a second bit from the first one. Assuming that, for example pattern $\underline{1}$000$\underline{1}$ will be identified as the most frequent for example, then the original coefficient will be replaced with 01000$\underline{1}$00 and pattern $\underline{1}$000$\underline{1}$ will be added to the coefficient set as a new coefficient.

Bit pattern	**Frequency**
10$\underline{1}$	1
1000$\underline{1}$	1
100000$\underline{1}$	1
$\underline{1}$0$\underline{1}$	1
$\underline{1}$000$\underline{1}$	1

Table 2. 2 non-zero-bit patterns of coefficient 010$\underline{1}$0$\underline{1}$0$\underline{1}$

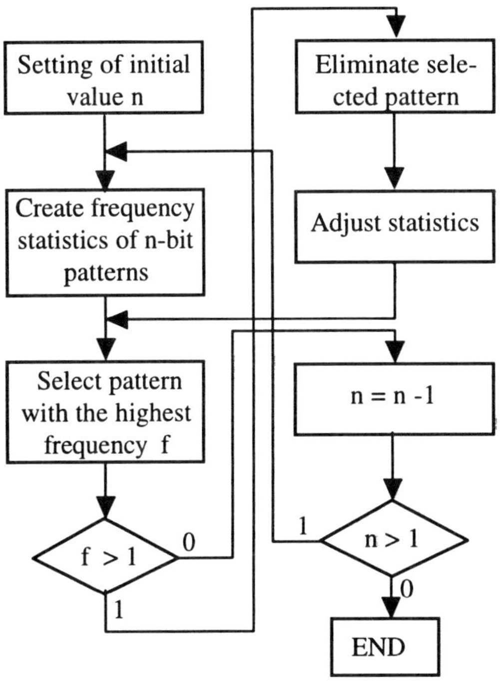

Figure 4. Optimization algorithm block-diagram

IV.2 Pattern Selection

In the case of patterns with the same frequency, some decision criterion is necessary to make a selection. We have chosen a very simple one. Patterns with more bits must be implemented in adder-subtractor structures with a bigger wordlength, and thus result in a larger area. If two patterns have the same frequency, the smallest pattern is chosen for further processing. For example, pattern 101 is preferred over pattern 1001. In a case of two bit patterns of equal length, one additional criterion is added: the preference of adders over subtractors. This is because an adder structure is smaller than a subtractor of the same wordlength. For example, pattern 101 will be selected rather than 10$\underline{1}$.

IV.3 Searching Algorithm

Because an exhaustive search is performed, the selection of an appropriate algorithm has crucial importance. We have decided to use a binary tree structure to hold the frequency statistics of patterns. As keys in the tree structure bit-patterns are used.

Initially, the pattern statistics are constructed as shown in Figure 5. For each coefficient, a local tree is constructed. After all patterns for that coefficient are evaluated, the global statistics tree holding information on the complete coefficient set, is updated.

After elimination of a pattern, the frequency of other patterns can also change, and therefore the frequency statistics must be re-evaluated after each elimination. Since the creation of a new global tree after each pattern elimination is unacceptable, an alternate method of adjusting the statistics is used (Figure 6). The idea is to use a local tree with difference frequency statistics for every changed coefficient that is processed. This local tree holds the information about changes in frequencies for the different subexpressions, after pattern elimination. The global tree can then be updated with this difference information after common subexpression removal. For example, in Figure 6, the frequency of pattern 101 frequency has changed from 2 to 1, due to elimination of the subexpression 1000001 out of the coefficient 0.10101010, so the difference frequency of pattern 101 is -1 (one less) thus the global frequency of this pattern must be also decremented. By backannotating the global tree after common subexpression removal, the global statistics have to be created only once at the beginning.

IV.4 Optimality of the solution

We have introduced an exhaustive search mechanism in the selection of bit patterns *per design iteration*. This is not equivalent to the exhaustive search over the complete design space considering *all design iterations at once*. It can be shown however, that the selected strategy leads to the optimal result with probability higher than 0.5. In other words, even if the selected strategy is not guaranteed to be optimal, we can guarantee a good performance on the average.

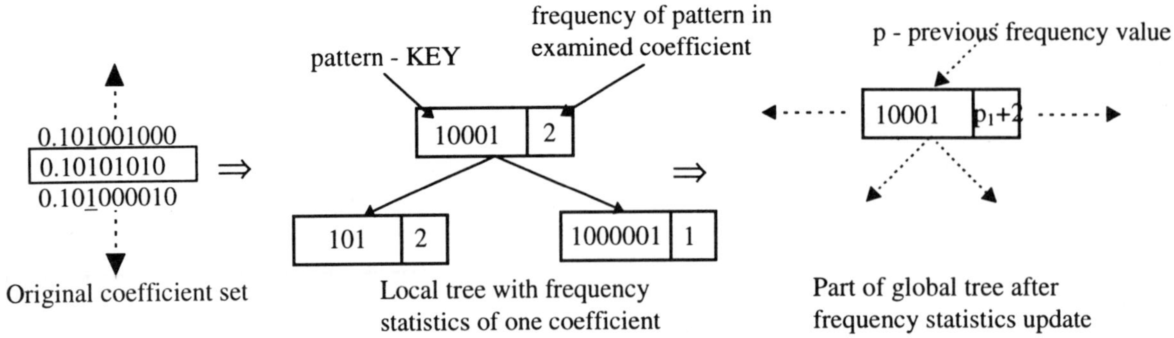

Figure 5. Creation of a new frequency statictics

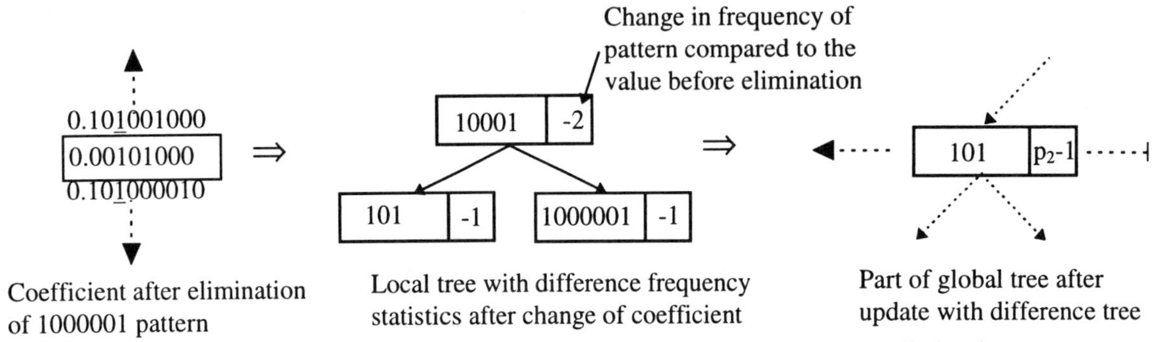

Figure 6. Frequency statistics re-evaluation after 1000001 patern elimination

	8 bit CSD coeff.			**12 bit CSD coeff.**			**16 bit CSD coeff.**		
# of taps	**init**	**opt**	**R**	**init**	**opt**	**R**	**init**	**opt**	**R**
16	**23**	**12**	**1.9**	**41**	**18**	**2.3**	**58**	**22**	**2.6**
32	**40**	**19**	**2.1**	**77**	**30**	**2.6**	**105**	**34**	**3.1**
64	**64**	**24**	**2.7**	**131**	**46**	**2.8**	**181**	**48**	**3.8**
128	**81**	**26**	**3.1**	**178**	**52**	**3.4**	**250**	**56**	**4.5**
256	**107**	**35**	**3.1**	**239**	**68**	**3.5**	**331**	**68**	**4.9**

Table 3. Experimental results

V. Design Examples

In this section, we present the results that are obtained using the proposed algorithm. First, the performance of the approach is discussed, and next, a comparison with the work of other authors is made.

V.1 Performance

The CSD conversion as well as the CSE algorithm has been developed in C. In the first example, three sets of FIR filters with CSD coefficients of 8,12 and 16 bits chosen at random were processed. The number of taps was 16, 32, 64, 128 and 256 for each set. The results are given in Table 3 in terms of the number of adders-subtractors necessary to implement the structure of the multiplier block (See Figure 3(a). We give the number of add/sub before (init) and after optimization (opt) as well as the optimization ratio R=init/opt. The results demonstrate, that the number of adders-subtractors for this random set of coefficients can be reduced by an average factor of 3.1 and the whole optimization works the better, the more coefficients are processed. Execution times of CSE algorithm for all examples were less than 1 second on a HP700 workstation.

In the second example, a Root Raised Cosine 64-tap FIR filter with impulse and frequency responses shown in Figures 7 and 8 respectively,10 bit CSD coefficients, 10 bit input and 16 bit output bus was optimized. After optimization, behavioral VHDL code was generated and synthesized using SYNOPSYS. Synthesis results are in Table 4. The area value is given as a sum of combinatorial and sequential area.

Area and critical path (t_{crit}) values are evaluated for Alcatel-Mietec 0.7um 5V standard cell technology.

	Unoptimized	Optimized
Area-com.[mm^2]	5.181	4.4804
Area-seq [mm^2]	3.9415	3.9415
Area-tot. [mm^2]	9.1224	8.4219
t_{crit}[ns]	41.1	41.46

Table 4. Synthesis results

V.2 Related Work Comparison

Optimization of FIR filters by means of removal of redundant hardware was already reported by several authors [5 - 8]. A comparison was made based on results presented for random coefficients.
We compare the average optimization ratios R.
. Table 5 shows that the average optimization ratio of 3.1 that we obtain is clearly better than that what was presented before.

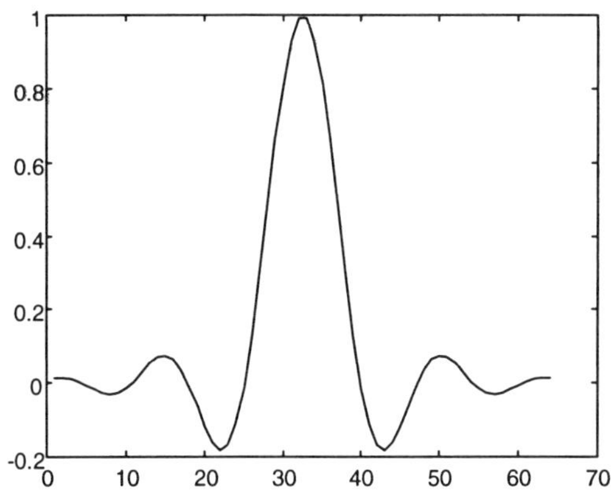

Figure 7. Filter impulse response

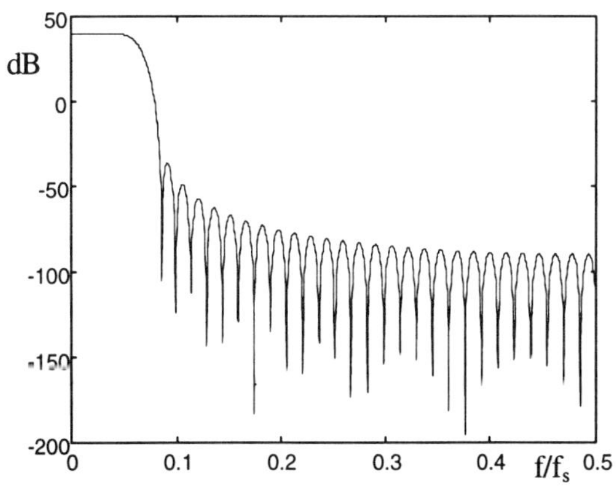

Figure 8. Filter frequency response

Author	R
Potkonjak [6]	1.4
Hartley [7]	2.1
Mehendale [5]	2.2
Demptser [8]	2.7
This work	3.1

Table 5. Related work results comparison

VI. Conclusion

In this paper, we presented an approach to the Multiple Constant Multiplication problem defined in [6] by exhaustive search. The results on our experimental filter set showed a significant reduction of hardware necessary in the multiplier block of a transposed form FIR filter with better performance than previously published papers. By including behavioral code generation as a postprocessing step of the optimization program, a tedious design subtask of broadband modem development has been automated and simplified.

Acknowledgment

This work was supported by COPERNICUS project SISPAS CP94-0223.

References

[1] Digital Broadcasting Systems for Television, ETS 300 429, ETSI, 1994.

[2] Digital Audio Visual Council, "Lower Layer Protocols And Physical Interfaces", DAVIC 1.0 Specification, Part 08, 1996.

[3] K. Hwang, "Computer Arithmetic", Jhon Wiley, New York, 1979.

[4] H. Samueli, "The Design of Multiplierless Digital Data Transmission Filters with Powers-of-two Coefficients", Proc. SBT/IEEE Int. Telecomm.Symp. September 1992, pp. 425 - 429

[5] Mehendale M., Sherlekar S.D., Venkatesh G., "Synthesis of Multiplier-less FIR Filters with Minimum Number of Additions", IEEE Transactions on computer aided design 1995.

[6] Potkonjak M., Srivastava M.B., Chandrakasan P.A., "Multiple Constant Multiplications: Efficient and Versatile Framework and Algorithms for Exploring Common Subexpression Elimination", IEEE Transactions on Computer-aided Design of Integrated Circuits and Systems, Vol 15. No. 2. Feb. 1996

[7] Hartley R.I., "Subexpression Sharing in Filters Using Canonic Signed Digit Multipliers", IEEE Transactions on Circuits and Systems-II: Analog and Digital Signal Processing, Vol. 43. No 10. Oct. 1996

[8] A.G. Demptser, M.D. Macleod, "Use of Minimum-Adder Multiplier Blocks in FIR Digital Filters", IEEE Transactions on Circuits and Systems-II:Analog and Digital Signal Processing, Vol. 42, No 9, Sept. 1995.

Port Calling: A Transformation for Reducing I/O during Multi-Package Functional Partitioning

Frank Vahid
Department of Computer Science
University of California, Riverside, CA 92521
vahid@cs.ucr.edu, www.cs.ucr.edu/~vahid

Abstract

Partitioning a system among multiple input and output pin (I/O) limited packages is a widely researched and hard to solve problem. We previously described a new approach yielding large improvements, which partitioned functions rather than structure, and which used a single bus for all inter-package data transfer. In this paper, we describe an extension permitting arbitrary distribution of I/O among the packages, and highlight experiments demonstrating even further I/O reductions as well as surprisingly improved performance, with nearly no penalty.

1 Introduction

Systems often must be partitioned among multiple packages. For example, an embedded system might use a software micro-controller package for those functions that need not run very fast or that are likely to change, and a hardware part for functions requiring high-speed execution. The hardware part might in turn consist of a number of FPGA packages, since one FPGA might not have enough gates or I/O to implement all the hardware functions. Even with today's increasing package capacities leading to single-chip solutions for entire systems, such a chip itself may still include one or more processor cores, as well as numerous hardware blocks. Logic emulation is another case requiring system partitioning among numerous FPGA's.

The multi-package partitioning problem has been widely researched, but has been hard to solve well. A good survey of the problem is found in [1]. Much of the research focuses on the fact that multi-package partitioning is I/O driven, meaning the number of I/O pins available on a package is the hardest constraint to satisfy, leading to underutilized gates and hence more packages than gate count alone implies. More packages can lead to larger systems, more power consumption, and reduced performance due to more inter-package signal crossings. Recent work [2] significantly reduced I/O for logic emulation, by time-multiplexing inter-package signal transfers over physical I/O, using multiple clocks and distributed controllers. Like most previous approaches, this approach performs structural partitioning, in which one first designs a system's structure (register-transfer components and gates), and then partitions the structure among packages.

We demonstrated significant advantages of a new and different approach, functional partitioning, over structural partitioning [3]. In functional partitioning, we take advantage of the recent trend of designers first specifying a system's functionality using a program-like language, such as VHDL, Verilog, or C. We partition this program among packages, using SpecSyn estimators developed by UC Irvine [4] to guide the partitioning process. After partitioning, structure may be designed separately for each package. The advantages included greatly reduced I/O (often 50%), improved performance, and reduced synthesis runtimes (often by an order of magnitude). Such advantages were predicted by some researchers for many years [5, 6, 7, 8, 9]. In addition, since programs are being partitioned rather than gates, the approach supports hardware-software partitioning, unlike structural partitioning which is for hardware only. Many hardware-software functional partitioning approaches have recently been proposed [10, 11, 12, 13, 14, 15, 16, 17].

In [18], we described an approach to implementing inter-package data transfers, using the FunctionBus. In this paper, we utilize the FunctionBus feature of fixed inter-package I/O size to develop an improved approach. This new approach uses a transformation we refer to as *port calling*, which enables us to almost arbitrarily distribute external port I/O among the packages, permitting better balancing of I/O, and often even improving performance and reducing total I/O, with nearly no penalty. We first provide background on the representation used and on the FunctionBus, introduce the port-calling transformation, describe the port-calling functions that are used, and summarize

```
void F()
{                        void G(int a, int& b)
  a = Q / 32;            {
  G(a, b);                 b = 1;
  H(b);                    for (int i=0; i<100; i++)
}                          {
                             b = b + a*R;
void H(int b)                H(b);
{                          }
  S = b * 100;           }
  P = d * 32;
}
```

Fig. 1: Example specification.

experiments demonstrating the significant improvements obtained.

2 Background

In this section, we briefly describe the SLIF representation and the FunctionBus, which are used by our technique. In our approach, a program (VHDL or C) is first translated to SLIF (System-Level Intermediate Format [19]). The SLIF, similar to a call-graph used in software profiling, is a directed graph, where each node represents a coarse-grained function or variable, and each edge represents an access by a function to another function or variable. The edge direction indicates the accessor and accessee, not the direction of data flow, which can occur in either or both directions over the same edge. For example, Figure 1 shows a (trivial) functional specification program written in C, consisting of three functions F, G and H, which call each other, and which access four external 16-bit ports P, Q, R and S. Figure 5(a) contains the SLIF for this program. Each function becomes a node, and each function call and port access becomes one edge.

During a pre-estimation phase, each SLIF edge is annotated with estimation and profiling information, such as the number of data bits for each transfer (function parameters or variable data) and the access frequency. Each SLIF node is also annotated with information, such as the execution time and size on various package types. Only external-port bit-size annotations are shown in the example.

During partitioning, online estimation combines these annotations, using non-trivial equations, to obtain fast yet adequately accurate estimates of design metrics, such as size (gates or bytes), I/O, and execution time. For example, the execution-time equation considers the time for a procedure to execute on its current component, plus the time to transfer data to and from any accessed procedures and variables (which depends heavily on the locality of those objects in the current partition), plus the execution time for accessed procedures. The cost function guiding partitioning is a weighted-sum of normalized constraint violations. The partitioning itself consists of moving SLIF nodes among packages, where each package implements a custom or standard processor, and is thus performed at the granularity of procedures and large variables (though procedures can be inlined or exlined to adjust the granularity). It is achieved by using any of several heuristics (such as simulated annealing, a modified Kernighan/Lin heuristic, clustering or greedy improvement) or by manually moving objects. For further details on the partitioning approach, the reader is referred to [19, 4, 12, 20].

In addition to using the SLIF representation, our approach uses the FunctionBus for inter-part communication [18]. In contrast, many earlier multi-package functional partitioning approaches used a cut-edges approach to I/O implementation. In such an approach, a graph's nodes, representing functions of varying granularities, are partitioned among parts. When an edge, which represents data, is cut (i.e., crosses between two parts), a unique set of I/O pins is used to transfer the data. However, observing that a large process often must be partitioned among parts, and that a process' functions can or do execute sequentially, we used a data transfer approach using what was called a FunctionBus. The bus consists of one address valid line, one data valid line, and a set of multiplexed address/data lines of a designer-chosen width, as shown in Figure 2. The bus works as follows. Each function is assigned a unique address. Whenever a function A calls a function B on another part, A first places B's address on the bus and asserts the address line, as shown in the timing diagram of Figure 3(a). If B is called by more than one function, then A must also send its own address as a return address. A then places any input parameters that must be passed to B on the bus, in chunks if the parameter size exceeds the bus size, asserting the data valid line for each chunk, and then A suspends. B detects its address, receives the parameter data, executes (perhaps calling other functions), and then returns to A by sending A's address over the bus followed by any output parameters. A detects its address, receives any parameter data, and resumes execution. Figure 4 provides example send and receive C routines for transferring long integer data over a FunctionBus with 8 address/data lines.

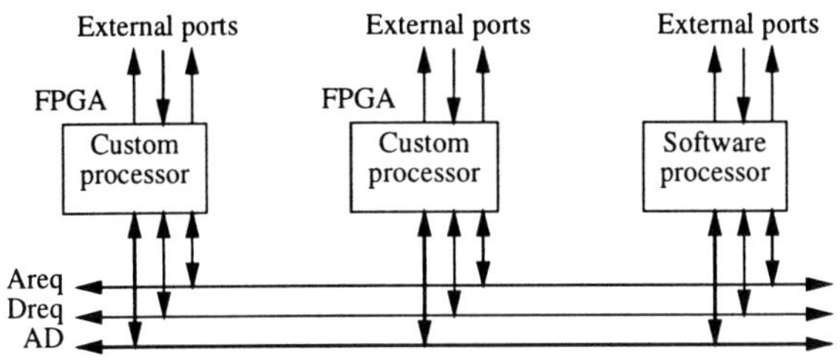

Fig. 2: FunctionBus architecture.

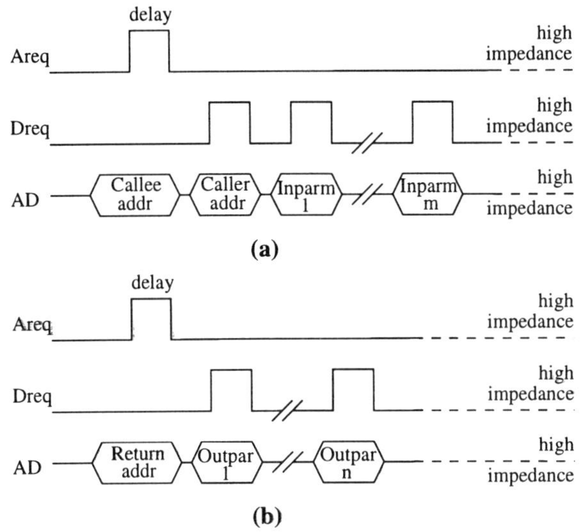

Fig. 3: Timing diagrams: (a) function call, (b) function return.

We demonstrated that the FunctionBus yielded even further reductions in I/O over the already large reductions of functional partitioning over structural partitioning. The multiple functions of a process on a given part require only one process, which looks for an address of any of its functions, and then branches to the appropriate one. In addition, the FunctionBus could be used for either hardware partitioning, or for hardware/software partitioning where the software component had parallel I/O (e.g., a micro-controller having several N-bit ports to which it can read or write, independent of memory accesses).

3 Port-calling transformation

In a FunctionBus approach, the number of I/O pins required for communication among packages is fixed (and typically small, like 18 or 34 pins). Hence, the only variation in a package's I/O comes from the I/O connected to external ports, needed by the functions on that package. The port-calling transformation will allow us to redistribute such ex-

```
void FB_SendLong(
    byte fb_addr;
    long fb_data )
{
  // Send the address
  FB_AD     = fb_addr;
  FB_Areq   = 1;
  FB_Delay();
  FB_Areq   = 0;

  // Send the data
  for (i=0; i<4; i++)
  {
    FB_AD =fb_data>>(8i)
    FB_Dreq   = 1;
    FB_Delay();
    FB_Dreq   = 0;
  }

  // Release the bus
  FB_AD = Z;
  FB_Areq = FB_Dreq = Z;
}
```

```
char FB_RecLong(
    byte fb_addr )
{
  long fb_data;

  // Wait for the address
  while ( ! ( FB_Areq &&
              FB_AD==fb_addr
        ) );

  // Receive the data
  for (i=0; i<4; i++)
  {
    while (! FB_Dreq);
    fb_data
      = (fb_data<<8)|FB_AD;
    while (FB_Dreq);
  }

  // Return the data
  return(fb_data);
}
```

Fig. 4: FunctionBus long-data transfer routines.

ternal port I/O to packages other than the accessing function's package.

The transformation consists of introducing a new function, called a port-call function, in between the original accessor function and the port itself. This function may, upon being called by the accessor function, read the port or write the port (as will be discussed further in the next section) on behalf of that function. Thus, from the accessor's perspective, accessing the port has been replaced by a function call.

In SLIF, a port-call function is represented as any other function, i.e., a node. This node can be partitioned among packages just like any other function. If this node is separated from its accessor function, any data transfer will take place over the existing FunctionBus; since the I/O for the FunctionBus already exists and is fixed, such data transfer does not require any additional inter-package I/O. Since a port-call node has extremely simple contents and hence when implemented will not contribute noticeably to a package's size, it can be partitioned to nearly any package. Hence, we see that introducing port-call nodes, in conjunction with the FunctionBus, yields the ability to freely distribute I/O among packages, at the possible expense of a few extra clocks cycles required to pass the data along from the port-call function to the accessor function.

For example, Figure 5(b) provides a simple example of a partitioned SLIF with several port accesses. Assuming a FunctionBus size of 18, the total I/O for the part on the right would be 66, since 48 I/O pins are required by the part's functions that

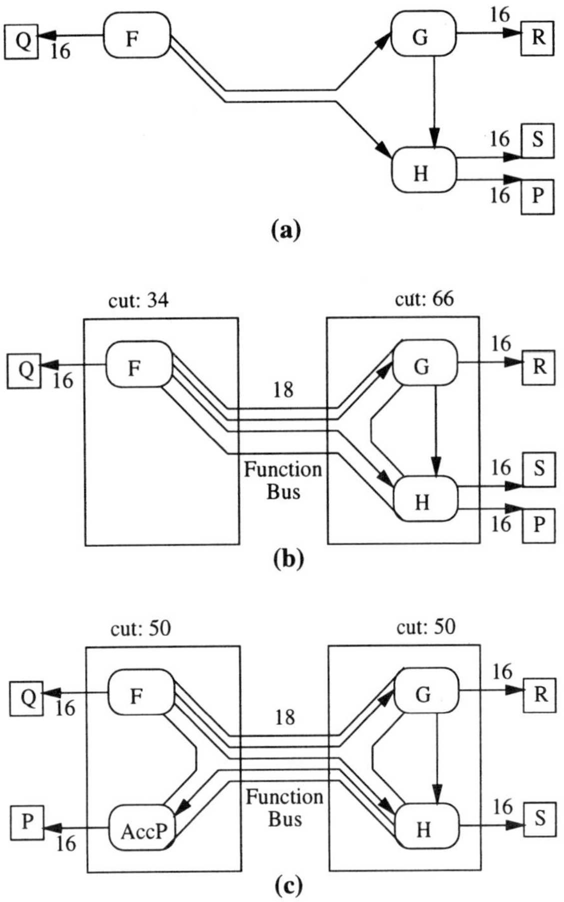

Fig. 5: Port calling example: (a) Original SLIF, (b) Partitioned SLIF using FunctionBus, (c) After port-calling transformation.

access three 16-bit external ports. However, notice that the left part only requires 34 I/O. There is a significant imbalance in I/O. Though the total I/O is 100, we would not be able to use two 50-pin packages, but instead would have to use a more expensive and larger package on the right with at least 66 pins.

If, however, we insert a port-call function *AccP* in between function *H* and port *P*, as shown in Figure 5(c), we can repartition the SLIF to achieve a perfect balance of 50 I/O for each part. *H* now must call *AccP*, which in turn accesses *P* directly. The write data is transferred from *H* to *AccP* over the FunctionBus.

We must decide which port accesses should have port-call nodes introduced to achieve improvements like that illustrated in Figure 5. Predicting those accesses that, when replaced by port-call nodes, would

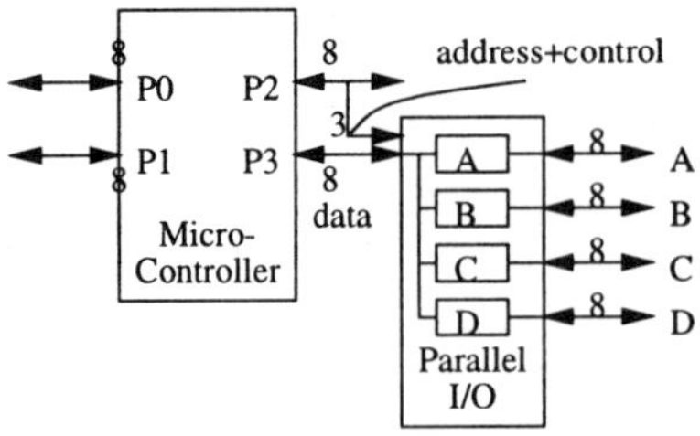

Fig. 6: Extended parallel I/O.

lead to such improvements, is a difficult task. We note, though, that the number of external ports is such that we can introduce a port-call node for every port access without much problem. Thus, our approach is to:

1. *Transform* the SLIF by introducing port-call nodes for every port access,
2. *Partition* the SLIF using existing heuristics, and
3. *Untransform* the SLIF by eliminating each port-call node that appears on the same part as its accessor.

The *untransform* step is necessary to eliminate unnecessary port-call nodes, so that only those nodes needed to distribute I/O to another part remain.

Note that port-calling is a generalization of the commonly used technique of extended parallel I/O. For example, consider Figure 6. A micro-controller with limited ports must interface to four 8-bit external ports A, B, C and D, using just one 8-bit port P3 and a few bits of P2. A common solution to this problem is to introduce a parallel I/O (PIO) chip, which multiplexes the four external ports over the single 8-bit data port, or demultiplexes the 8-bit data port to the four external ports, depending on its input address and control lines. The bus between the micro-controller and PIO chip is akin to the FunctionBus, and the control internal to the PIO is essentially equivalent to port-call functionality. Port-calling is more general since we can move the functionality to chips other than just PIO chips, such as to an FPGA or even another micro-controller.

4 Port-call functions

After the SLIF is partitioned, new program-like descriptions must be generated for each package. For a process *A* from the original specification, each

```
datatype PortCallRead()
{
   return(P);
}
void PortCallWrite(datatype d)
{
   P = d;
}
datatype PortCallReadOrWrite(datatype d, bit read)
{
  if (read)
    return(P)
  else
    {P = d; return(0);}
}
```

Fig. 7: Port-call functions.

package will have its own process for A. This process will consist of detecting one of its function's addresses on the FunctionBus, capturing any input parameters from the bus, calling the function, and then returning by placing a return address and any output parameters on the bus. A call to a function on another part is replaced by FunctionBus call and return routines, as shown in Figure 4. Port-call functions are treated as any other function, requiring no special treatment. We thus need only describe the contents of such functions here; partitioning and subsequent FunctionBus routine insertion will take care of the communication between the port-call function and the accessor function.

The port-call functions for accessors that read, write, or both read and write a port are shown in Figure 7. Note that each is trivial to implement, so could be moved freely among parts.

5 Experiments

We implemented SLIF transformations to insert port-call functions for every port and to delete port-call functions accessed only by functions on their own parts. We incorporated these transformations with existing tools that convert a VHDL specification to SLIF (using SpecSyn [12] estimators to annotate the SLIF with size, I/O, and execution information), and that partition the SLIF using standard heuristics (we used simulated annealing here) making use of the FunctionBus. We then applied two-way partitioning with and without port-calling on five examples: an answering machine controller *ans*, and Ethernet coprocessor *ether*, an fuzzy-logic controller *fuzzy*, an interactive TV processor *itv*, and a microwave transmitter controller *mwt*, each consisting of a few hundred lines of VHDL algorithmic code. Partitioning was achieved using simulated annealing, with a cost function that sought to

Example	Without Port Calling					With Port Calling				
	Size1	Size2	IO1	IO2	Time	Size1	Size2	IO1	IO2	Time
ans	6140	6491	85	98	44	6164	6494	32	35	44
ether	13326	11343	39	48	194	13478	11202	38	37	194
fuzzy	51899	59802	26	34	10888	58511	53193	34	26	8356
itv	51649	101483	91	61	10049	53288	99872	70	57	9653
mwt	4511	5447	40	32	799	4877	5095	35	23	791

Fig. 8: Partitioning without and with port calling.

minimize I/O and execution time while balancing part sizes. Details of the partitioning and estimation system are extensive and have been described elsewhere [12, 4].

Results are summarized in Figure 8. The table shows the size in gates for each part, the I/O for each part, and the execution time of the example, including communication time over the FunctionBus. Port calling yielded significant improvements. Note that not only is the maximum required I/O usually reduced, but that in most cases, total I/O is also reduced and performance is actually improved; these improvements come at the cost of negligible total size increase in some cases.

The improvements can be better seen in Figure 9. The first chart shows the maximum I/O required on either part without and with port calling. Note the reductions, especially in the case of *ans*. The second chart shows the reductions in total I/O, which is the sum of the I/O of both parts. The third chart shows the reductions in execution time. Note that we developed port-calling in order to achieve improvements in the maximum I/O metric; we would have been satisfied with the same total I/O and a slight penalty in performance (due to extra cycles for transferring port data over the FunctionBus). However, those other metrics were actually *improved*. Such improvement can be explained by noting that the maximum I/O is a difficult metric to satisfy, and it can dominate the cost function and hence steer the partitioning heuristics. By introducing port-calling, which allows for arbitrary I/O distribution, this pressure is greatly relieved, and hence the cost function is driven to a greater extent by the other metrics, so the partitioning heuristics can seek to improve those metrics; hence, we see substantial improvements in those metrics.

Future work includes investigating the reduction in power that can be obtained using a FunctionBus-based approach to functional partitioning, developing heuristics for partitioning, and developing libraries for communication among the various hardware and software parts after partitioning.

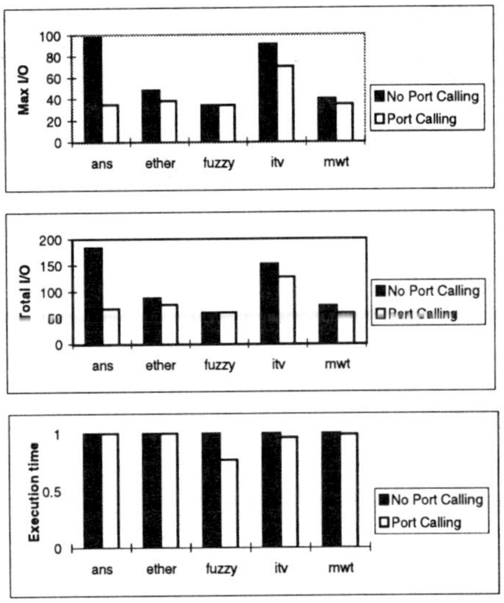

Fig. 9: Partitioning without and with port calling.

6 Conclusions

We have introduced an approach allowing for nearly arbitrary distribution of I/O among packages during partitioning. The approach extends functional partitioning with a port-calling transformation, using the FunctionBus for inter-package communication. The ability to arbitrarily distribute I/O among packages leads to smaller maximum I/O requirements and hence can reduce package cost or quantity. In addition, by easing this previously burdensome metric, partitioning heuristics can focus on other metrics, like performance and total I/O, and hence we see even further improvements in those metrics. Such improvements are important for multi-package partitioning, as well as for partitioning among blocks within a single large ASIC, which may become necessary to perform hardware-software partitioning or to reduce design complexity, routing complexity, power, and synthesis runtimes. Thus, port-calling in conjunction with the FunctionBus can significantly improve increasingly-important functional partitioning environments.

References

[1] F. Johannes, "Partitioning of VLSI circuits and systems," in *Proceedings of the Design Automation Conference*, 1996.

[2] R. Tessier, J. Babb, M. Dahl, S. Hanono, and A. Agarwal, "The virtual wires emulation system: A gate-efficient asic prototyping environment," in *Proceedings of the International Symposium on Field-Programmable Gate Arrays (FPGA)*, 1994.

[3] F. Vahid, T. Le, and Y. Hsu, "A comparison of functional and structural partitioning," in *International Symposium on System Synthesis*, pp. 121–126, 1996.

[4] D. Gajski, F. Vahid, S. Narayan, and J. Gong, "Specsyn: An environment supporting the specify-explore-refine paradigm for hardware/software system design," *IEEE Transactions on Very Large Scale Integration Systems*, p. to appear, 1997.

[5] E. Lagnese and D. Thomas, "Architectural partitioning for system level synthesis of integrated circuits," *IEEE Transactions on Computer-Aided Design*, vol. 10, pp. 847–860, July 1991.

[6] R. Gupta and G. DeMicheli, "Partitioning of functional models of synchronous digital systems," in *Proceedings of the International Conference on Computer-Aided Design*, pp. 216–219, 1990.

[7] K. Kucukcakar and A. Parker, "CHOP: A constraint-driven system-level partitioner," in *Proceedings of the Design Automation Conference*, pp. 514–519, 1991.

[8] Y. Chen, Y. Hsu, and C. King, "MULTIPAR: Behavioral partition for synthesizing multiprocessor architectures," *IEEE Transactions on Very Large Scale Integration Systems*, vol. 2, pp. 21–32, March 1994.

[9] C. Gebotys, "An optimization approach to the synthesis of multichip architectures," *IEEE Transactions on Very Large Scale Integration Systems*, vol. 2, no. 1, pp. 11–20, 1994.

[10] R. Gupta and G. DeMicheli, "Hardware-software cosynthesis for digital systems," in *IEEE Design & Test of Computers*, pp. 29–41, October 1993.

[11] R. Ernst, J. Henkel, and T. Benner, "Hardware-software cosynthesis for microcontrollers," in *IEEE Design & Test of Computers*, pp. 64–75, December 1994.

[12] D. Gajski, F. Vahid, S. Narayan, and J. Gong, *Specification and design of embedded systems*. New Jersey: Prentice Hall, 1994.

[13] X. Xiong, E. Barros, and W. Rosentiel, "A method for partitioning UNITY language in hardware and software," in *Proceedings of the European Design Automation Conference (EuroDAC)*, 1994.

[14] A. Kalavade and E. Lee, "A global criticality/local phase driven algorithm for the constrained hardware/software partitioning problem," in *International Workshop on Hardware-Software Co-Design*, pp. 42–48, 1994.

[15] P. Eles, Z. Peng, and A. Doboli, "VHDL system-level specification and partitioning in a hardware/software co-synthesis environment," in *International Workshop on Hardware-Software Co-Design*, pp. 49–55, 1992.

[16] P. Knudsen and J. Madsen, "PACE: A dynamic programming algorithm for hardware/software partitioning," in *International Workshop on Hardware-Software Co-Design*, pp. 85–92, 1996.

[17] A. Balboni, W. Fornaciari, and D. Sciuto, "Partitioning and exploration strategies in the tosca co-design flow," in *International Workshop on Hardware-Software Co-Design*, pp. 62–69, 1993.

[18] F. Vahid, "I/O and performance tradeoffs with the functionbus during multi-FPGA partitioning," in *Proceedings of the International Symposium on Field-Programmable Gate Arrays (FPGA)*, pp. 27–34, 1997.

[19] F. Vahid and D. Gajski, "SLIF: A specification-level intermediate format for system design," in *Proceedings of the European Design and Test Conference (EDTC)*, pp. 185–189, 1995.

[20] F. Vahid and D. Gajski, "Incremental hardware estimation during hardware/software functional partitioning," *IEEE Transactions on Very Large Scale Integration Systems*, vol. 3, no. 3, pp. 459–464, 1995.

A Scheduling and Pipelining Algorithm for Hardware/Software Systems

Smita Bakshi[†]
Dept. of Electrical & Computer Engineering
University of California, Davis
Davis, CA 95616
bakshi@ece.ucdavis.edu

Daniel D. Gajski
Dept. of Information & Computer Science
University of California, Irvine
Irvine, CA 92697-3425
gajski@ics.uci.edu

Abstract

Given a hardware/software partitioned specification and an allocation (number and type) of processors, we present an algorithm to (1) map each of the software behaviors (or tasks) to processors, (2) pipeline the system specification, and (3) schedule the behaviors in each pipe stage, amongst selected hardware components and processors, so as to satisfy a throughput constraint at minimal hardware cost. Thus, to achieve high performance, not only are critical tasks implemented as pipelined hardware architectures, but the system is also divided into concurrently executing stages. Furthermore, to offset the cost of this increased concurrency, non-critical sections are implemented on processors or as cheaper hardware blocks. Our experiments demonstrate the feasibility of our approach and the necessity of system pipelining in high performance design.

1 Introduction

Digital system design, especially within signal and image processing, is an immensely complex task and requires the fine-tuning and balancing of a number of different parameters. Some of the design parameters that have a large impact on the final design are the number and types of components (such as ASICs, general purpose processors, FPGAs) used in the design, the interconnection and communication mechanisms for these components, and the manner in which the different tasks within the system are partitioned, scheduled and pipelined amongst these components. Partitioning refers to dividing tasks amongst components such that critical tasks are implemented on faster components, while the less critical tasks are implemented on slower and cheaper components. Pipelining refers to dividing the tasks into concurrently executing stages, to increase the effective parallelism, and hence performance of the design. Finally, scheduling refers to deciding the sequentiality or the order of tasks within a pipe stage and on a component.

In this paper, we present an algorithm for scheduling and pipelining a specification amongst selected hardware and software components. The hardware components refer to RTL components such as adders, multipliers, ALUs, registers, and so on, while the software components refer to general-purpose processors such as the Pentium and PowerPC. This algorithm is part of a larger design flow in which the specification is first *spatially* partitioned into hardware and software tasks, and a hardware implementation is then determined for each of the hardware tasks. After the spatial partitioning, our algorithm performs temporal partitioning, that is, it divides the specification into pipe stages and time slots within the pipe stages.

The spatial partitioning allows the user to take advantage of the unique features offered by hardware and software components. Thus, critical sections may be placed in hardware and the less critical sections (or the ones that require programmability) may be implemented as software. Temporal partitioning, on the other hand, determines the level of concurrency in the design required to achieve a desired throughput constraint. Temporal partitioning is most needed for high performance applications, when the fastest implementation of critical behaviors still violates constraints. In this paper, we present an algorithm for temporal partitioning, given a spatially partitioned specification.

The next section describes related work in pipelining and hardware/software partitioning. Section 3 defines the problem and illustrates it with an example. We present the algorithm in Section 4, and experimental results along with conclusions in Sections 5 and 6, respectively.

2 Previous Work

Research in hardware/software partitioning has mainly concentrated on spatial partitioning [3] [6] [10] [14] wherein a specification is divided amongst multiple ASICs and processors to best satisfy constraints on area and/or performance. However, these tools have not considered the possibility of pipelining the system, at any level of granular-

† This work was performed while the author was at UC Irvine.
It was partially supported by the Semiconductor Research Corporation (Grant #93-DJ-146), and the authors gratefully acknowledge their support.

1080-1820/97 $10.00 © 1997 IEEE

ity, and hence fail to explore the design space within which most large DSP systems lie. Our work uses a very simple approach for partitioning, but by pipelining the system, it is able to explore this high-performance design space.

Several algorithms for scheduling and pipelining a data flow graph [9] or a control-data flow graph [8] [12] have been proposed in the past. These algorithms certainly form a basis for our pipelining and scheduling algorithm. An important difference, however, is that along with pipelining, we also determine a processor selection and binding for each of the software tasks. This places additional constraints on the algorithm, thereby increasing its complexity. Other differences are minor and arise mainly because of the different levels of granularity of pipelining. Our level of granularity is a task (a sequence of RTL operations), whereas for most algorithms it is an RTL operation.

We would also like to mention that the concurrency achieved due to pipelining is similar to the paradigm of concurrent tasks in synchronous data flow machines [11]. By pipelining and scheduling we are simply trying to balance the level of concurrency and sequentiality in a system, so as to minimize cost for a throughput constraint.

3. Problem definition

As mentioned in the introduction, our algorithm for scheduling and pipelining is part of a larger design flow [2]. To motivate our problem definition, we briefly describe this design flow. Given a system-level specification, hardware and processor libraries, and a throughput constraint, we first estimate the performance (number of clock cycles) of each task (or behavior) within the specification, on each of the available processors. Based on these estimates and the throughput constraint, we determine the hardware/software partition, that is, we associate a hardware or software type with each behavior. For all the hardware behaviors we then synthesize an implementation using components from the hardware library. This implementation may, in turn, be pipelined to satisfy constraints. We then obtain a processor allocation, and finally schedule and pipeline the hardware/software partitioned specification into pipe stages and amongst the selected processors. In this paper, we focus on this problem of scheduling and pipelining a partitioned specification.

The input specification is represented as a control flow graph, CFG(V,E), where vertices (V) represent tasks, and edges (E) represent control dependencies. Each task is represented by a sequence of VHDL statements which, in turn, are represented as a CDFG (control data flow graph).

For the given CFG, our aim is to determine the schedule and pipeline that will satisfy a throughput constraint, T. If we think of time as a grid $Time(x, y)$, where x is a continuum from $\{0 \cdots T\}$ representing time in ns, and $\{y = 1 \cdots \infty\}$ represents the number of pipe stages, then the problem may be defined as follows:

Given:

1. a partitioned control flow graph CFG(V,E) in which each $v \in V$ has been designated a hardware or software type
2. a processor allocation P (number and type of processors) selected from the processor library.
3. an execution time, T_v for every hardware node, $v \in V$
4. an execution time, T_{vp} for every software node, on each of the available processors, $p \in P$.
5. the throughput constraint, T.

Determine:

1. For every behavior $v \in V$, a start and finish point in the grid, (x_v, y_v) and (x'_v, y_v), where $x'_v > x_v$.
2. For every software behavior $v_s \in V$, a processor $p \in P$, which will be used to implement it.
3. For every processor $p \in P$, a utilization list containing pairs (x_{p1}, x_{p2}), indicating time intervals when the processor is utilized.

Such that:

1. $x'_v - x_v = T_v$, where T_v is the execution time in hardware or on a selected processor p_v, if v is a software behavior.
2. $x'_v < T, \forall v \in V$.
3. If $w \in Predecessor(v)$ is mapped to the start and finish grid points (x_w, y_w) and (x'_w, y_w), then $y_w \leq y_v$. Furthermore, if $y_w = y_v$, then $x_v > x'_w$.
4. If v is a software behavior mapped to processor p_v, then the time interval, (x_v, x'_v), does not overlap with existing time intervals in the utilization list of processor p_v.

In simpler words, we assign each behavior, v, to a pipe stage and to a time slot within a pipe stage such that predecessor behaviors of v finish their execution either in a previous stage, or in the same stage before the behavior v begins its execution. Furthermore, if v is a software behavior, then we have to make sure that the processor we select to execute it, is not used by any other behavior during the time interval that it is executing behavior v. This assignment of behaviors to pipe stages is done so as to (1) satisfy the throughput constraint, T, using the given processor allocation, P, and (2) to reduce the number of pipe stages (and therefore, the latency of the design and also the memory required to store data between pipe stages).

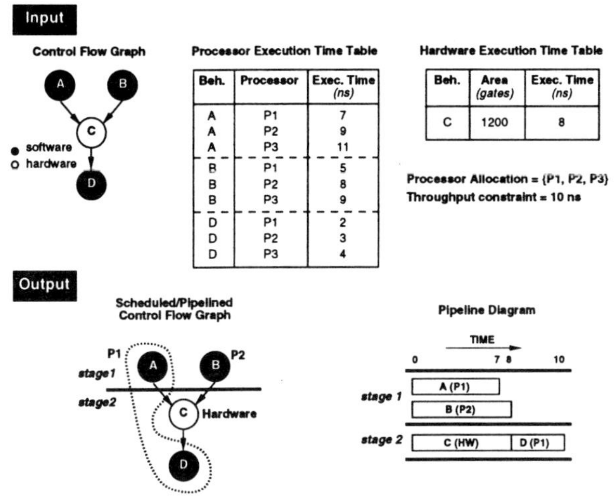

Figure 1. Algorithm inputs and outputs.

Our problem is illustrated with an example in Figure 1. The specification is given in terms of a control flow graph (CFG) of 4 behaviors (or tasks). Each vertex consists of VHDL statements representing some functionality of the specification. The behaviors in the CFG are partitioned into hardware and software. We are also given estimates of the execution time of each of the software behaviors (A, B and D) on each of the processors (P1, P2, P3 in this example). Similarly, we are given the area and execution time of the hardware implementation of each of the hardware behaviors (only behavior C in this example). Finally, we are given a processor allocation of one instance each of P1, P2 and P3, and a throughput constraint of 10 ns. This constraint indicates that a new sample of data will be available every 10 ns. Hence, the CFG may be partitioned into pipe stages of delay 10ns or less.

Our algorithm pipelines the CFG into 2 pipe stages, each of delay no more than 10 ns. It also determines the processor on which each of the software behaviors will execute (A on P1, B on P2 and D on P1). Finally, it determines the execution start and end times of all behaviors (nodes) within a pipe stage.

4 Algorithm overview

Our algorithm, outlined in Figure 2, is an extension of the well known list-scheduling algorithm [7]. We start by determining the longest completion time from each node till all output nodes, assuming that the fastest processor is used to execute a software node. This completion time gives the priority one node has over another during scheduling. The completion time of a node is a direct indication of its criticality, and hence, the higher the completion time, the higher its priority. In the example CFG in Figure 1 the completion times of nodes A, B, C and D are 17, 15, 10 and 2 respectively. Hence, the priority from highest to lowest is A, B, C and D.

```
1.  For every node in CFG, determine longest
    completion time from that node till any output node.
    Assign node priorities.
2.  Initialize the utilization list of all processors.
3.  Form a new ready list.
4.  Loop
5.      current_node = first node in ready list
6.      If (current_node is type software)
7.          Find processor and corresponding time slot
            that gives earliest completion time.
8.          If (no available processor)
9.              Exit (No feasible solution)
10.         Else
11.             Assign current_node to processor & time slot.
12.             Update utilization list of processor.
13.         End if
14.     Else if (current_node is type hardware)
15.         Assign current_node to earliest feasible time slot.
16.     End if
17.     Mark current_node as scheduled, remove from
        ready list, and update ready list.
18. Until (all nodes in CFG are scheduled).
```

Figure 2. Algorithm overview.

As mentioned in the previous section, with each processor, we associate a utilization list within which we store the durations in which the processor is being utilized to execute any of the software behaviors. In step 2, we reset the utilization list of all processors to empty. We then form a list of ready nodes, that is nodes whose predecessors have already been scheduled. These nodes are prioritized using the completion time priority function.

After forming the ready nodes' list and the utilization list for the processors, we find the "best time slot" for every node in the ready list, starting with the first (or highest priority) node. For nodes that are labeled as software, the "best time slot", of course depends on the processor selected to implement that node. Hence, a by-product of determining the best slot time is a processor selection for every software behavior. After a time slot (and processor) has been assigned for a node, we mark it as "scheduled", and then remove it from the ready list. Once removed, we update the ready list by checking to see if any of its successor's is now ready to be scheduled. If so, we add the successor(s) to the ready list. Steps 5 to 17 are repeated till all nodes in the *CFG* are scheduled.

The heart of the algorithm lies in determining the best time slot for each behavior (Steps 7 and 15), and a corresponding processor if the behavior is of software type. This is now explained in detail.

4.1 Determining the best time slot

Every node should ideally execute immediately after all its predecessors have completed execution, in the same pipe stage as it's last predecessor[1]. However, this may not be feasible for two reasons. Firstly, by scheduling it immediately after its predecessors, the throughput constraint may be violated, that is the pipe stage delay might exceed the constraint. Secondly, in the case of a software node, a processor for the desired duration may not be available. In both these situations it will then be necessary to schedule the node in the next pipe stage.

The procedure for determining the best time slot is outlined in Figure 3. The inputs to this procedure are the partially scheduled and pipelined CFG, the utilization list of all processors, and the node to be scheduled (*current_node*). We first select the maximum of all the pipe stages in which predecessors of *current_node* have been pipelined. We call this *max_pipestage*. We then determine the latest time at which predecessors **in max_pipestage** finish execution. This is given by the variable *max_preddelay*. These two quantities give us the earliest pipe stage and earliest time within that pipe stage that current_node may be scheduled in. Next, we check to see if *current_node* is in hardware or software. If it is a hardware node, we assign it to *max_pipestage* if it can complete its execution within the throughput constraint, else we assign it to *max_pipestage* + 1.

If it is a software node, we check the utilization lists of all processors and find the earliest completion time of current_node on all processors. At this point, we check two possibilities:

- the execution can begin after *max_preddelay* in *max_pipestage*
- the execution can begin at time 0 in *max_pipestage* + 1

From these two possibilities, we select the one which gives the earliest completion time. This may result in *current_node* being scheduled in *max_pipestage* or in *max_pipestage* + 1.

FindBestTimeSlot (CFG,utilization_lists,current_node)
Begin algorithm
max_pipestage = determine maximum pipe stage of all pred.
max_preddelay = determine latest completion time of all pred. executing in max_pipestage.
If (current_node is in hardware)
new_stagedelay = max_preddelay + delay(current_node)
If (new_stagedelay > ThruputC)
Assign current_node(pipe stage) = max_pipestage + 1
Assign current_node(start time) = 0.
Else
Assign current_node(pipe stage) = max_pipestage
Assign current_node(start time) = max_preddelay
End if
End if
If (current_node is in software)
comp_time1 = find earliest completion time on all processors with execution beginning after max_preddelay
comp_time2 = find earliest completion time on all processors with execution beginning after time 0
If (comp_time1 $\leq$ comp_time2)
Assign current_node(pipe stage) = max_pipestage
Assign current_node(start time) = max_preddelay
Else
Assign current_node(pipe stage) = max_pipestage + 1
Assign current_node(start time) = 0
End If
End If
End algorithm

Figure 3. Determining the best time slot and processor.

We explain the procedure for determining the best time slot by using the example in Figure 1. The algorithm starts by finding the longest delay from each node till any output node, assuming execution on the fastest processor. Thus, starting from the bottom of the *CFG*, node D has a delay of 2 *ns* (we'll drop the *ns* from now on) assuming execution on processor $P1$, node C has a delay of 10 (8+2), node B of 15 (10+5), and node A of 17 (10+7), giving node A the highest priority and node D the lowest. The initial prioritized ready list, then, consists of nodes A and B in that order.

We start by finding the best time slot for node A. We have a choice of three processors and corresponding three time slots: from 0 to 7 on processor P1, 0 to 9 on processor P2, and 0 to 11 on processor $P3$. This is indicated in the Completion Time Table in Figure 4(a). Each entry in the table gives the completion time and the pipe stage for a specific behavior on a specific processor. Note that the completion time table is not built before we start scheduling, but an entry for a node is appended to the list when the node is at the top of the ready list, next in line to be scheduled.

Of the three choices for node A we select processor $P1$ since that gives us the earliest completion time. Node A is thus scheduled in the first pipe stage, starting at time 0 and ending at time 7. The schedule as well as the utilization list of processor $P1$ is updated as shown in Figure 4(c) and (b), respectively. Next, the completion times for node B on all the processors is calculated, and processor $P2$ is selected

[1] The last predecessor is the predecessor that finishes its execution last *in the last pipe stage* amongst all predecessors. For instance, if Predecessor 1 finishes execution in stage 1, time 100 ns, and Predecessor 2 finishes execution in stage 2, time 50, then Predecessor 2 is the last one.

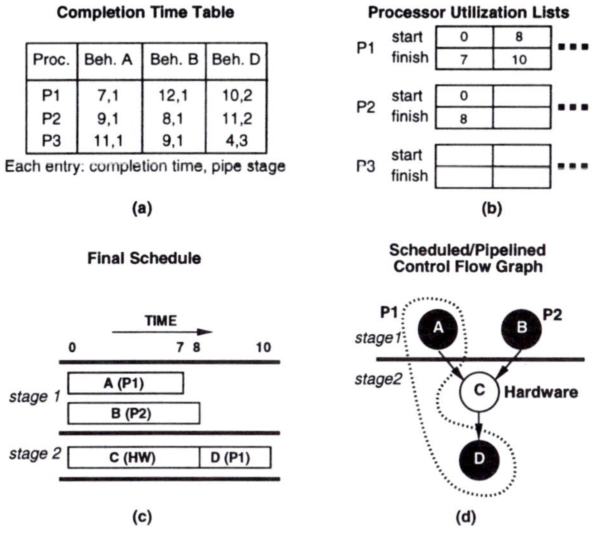

Figure 4. Determining the best time slot (and processor) for each node.

to be the best since it gives us the earliest completion time. Note that even though Processor $P1$ is faster and has a lower execution time than $P2$, it is utilized by A from 0 to 7, and thus it offers a completion time of 12 (7 + 5), which violates the throughput constraint of 10.

Next we come to node C, which is in hardware, and thus does not need to contend with any other node for its resources. The earliest we can schedule C is at time 0 in stage 2. If we pipeline it in stage 1, we will get a completion time of 16 (8 + 8), which is a violation of the throughput constraint.

Next, the ready list contains node D which cannot be scheduled prior to stage 2 since its predecessor has been scheduled in stage 2. The options for node D are: (1) on P1 from time 8 to 10 in stage 2, (2) on P2 from time 8 to 11 in stage 2, (3) on P3 from time 0 to 4 in stage 3. Of these four choices, we select the first, since it gives us the earliest completion time in the same pipe stage as its predecessor, node C. Though $P3$ gives us an earlier completion time, it introduces a third pipe. This is undesirable since one of the secondary goals of our algorithm is to minimize the total number of pipe stages.

The final schedule and pipeline is shown in Figure 4(c) and (d).

5 Experiments

We have implemented the scheduling and pipelining algorithm within the design flow outlined in Section 3. This work has been incorporated into SpecSyn [4], a system-level synthesis tool. Our experiments indicate the performance gains obtained by pipelining at the system level and at the behavioral level. They also demonstrate the extent and nature of design space explored. In an attempt to gauge the quality of our algorithm, we also compare the algorithmic design exploration against a manual one.

The experiments are conducted for the MPEG decoder II system and for the Volume system, an example of a medical instrument. Both, the MPEG and the Volume specification contain 14 behaviors or tasks (hence, 14 nodes in the *CFG*), though the MPEG specification is the larger of the two, with 1085 lines of code as opposed to 238 for the Volume system. For both examples we use a clock of 10 *ns*, a software library containing 6 processors (Intel's 8086, 8088, Pentium, Sun's SPARC, Motorola's 68000 and PowerPC), and a hardware library containing multiple implementations of RTL components such as adders, multipliers, and comparators. The processor and hardware execution time tables are obtained by using software [5] and hardware [1] estimators implemented within SpecSyn.

In order to explore the design space of both examples, we run our algorithm with a range of throughput constraints, from low to high. Our algorithm returns a set of designs ranging from all hardware, highly pipelined solutions, to ones containing processors and possibly custom hardware, and finally to the slowest ones that are implemented solely on a processor. Results for this algorithmic exploration are shown in Tables 1 and 3, in Figure 5. For the MPEG example, the exploration was also performed manually [13] (Table 2) For both examples, the tables indicate the number of **system-level** pipe stages and the processors and/or area required for different throughput values. Recall that system-level pipelining divides the CFG of tasks into pipe stages. This is as opposed to behavioral-level pipelining, in which each task in the CFG is further pipelined.

We make the following observations from the results in Tables 1 and 3:

- For the MPEG example, the fastest throughput attainable by our algorithm is 2980 *ns*. This design contained 12 pipe stages at the system-level. In addition, about 5 of the 14 behaviors were further pipelined into stages ranging in number from about 3 to 6. If we do not pipeline the design at the system-level, but still allow pipelining within behaviors, the fastest design we can obtain has a throughput of about 30000 ns, that is approximately 10 times that of the design obtained with system pipelining.

- For the Volume system, the fastest throughput attainable by our algorithm is 420 ns. However, without system-level pipelining, the throughput is approximately 2500 ns, which is about 6 times the highest throughput.

Both these observations indicate that in order to obtain high throughputs it is not sufficient to pipeline individual

MPEG System
Our Algorithm

Thruput (*ns*)	# Pipe Stages	Processors & FU + mem area (*gates*)
2980	12	462304
3840	12	395838
4480	11	376998
5760	12	Pentium x 2 335307
7680	12	Pentium x 3 224455
17810	10	Pentium x 3 152825
653970	1	Pentium x 1

TABLE 1

MPEG System
Manual Exploration

Thruput (*ns*)	# Pipe Stages	Processors & FU+mem+mux+reg+control area (*gates*)
–	–	–
4000	12	492953
4988	12	491692
5780	12	Pentium x 1 477114
7172	12	Pentium x 2 357470
17670	12	Pentium x 4 263167
661622	1	Pentium x 1

TABLE 2

Volume System
Our Algorithm

Thruput (*ns*)	# Pipe Stages	Processors & FU + mem area (*gates*)
420	6	6541
940	4	6231
1680	3	5715
3470	1	5715
8340	2	Sparc x 1 Pentium x 1 1497
17010	1	PowerPC x 1
52520	1	68020 x 1

TABLE 3

Figure 5. Designs explored by partitioning and pipelining the MPEG and Volume System.

behaviors. The pipelining needs to be extended to the system level such that the control flow graph of behaviors is further divided into concurrent stages, thereby reducing the throughput.

From these results, we can also obtain a comparison of the algorithmic and manual design process for the MPEG example (Tables 1 and 2). Note that in the manual designs, the hardware area is the area of the entire datapath and controller, while our algorithm just estimates the functional unit and the memory area. Hence, in general, the area of the manual design is higher than the estimated area. The results indicate that the design exploration conducted by our algorithm closely matches the manual exploration. Though the accuracy of our hardware implementations is not very high, its fidelity is extremely high. Also note that our algorithm obtained a faster design, than a designer could obtain manually. This is attributed to hierarchical pipelining (within a system, behavior, loop, and operation) which an algorithm performs easily, but is complex for designers to perform manually.

6 Conclusions

We have presented an algorithm to schedule and pipeline a hardware/software partitioned specification, given a processor allocation and a throughput constraint. Results have indicated the feasibility of our approach as well as the necessity of system and behavioral pipelining in order to obtain high data rates. However, our approach has several drawbacks, some of which we are currently addressing. Principal amongst these is the exclusion of communication overheads (in terms of area and delay) in our model and results. After this and other improvements, such as estimation of controller area, we expect our approach to offer a viable and a much faster alternative to manual design and exploration.

References

[1] S. Bakshi. *Hardware/Software Co-design for Pipelined Systems*. PhD thesis, University of California, Irvine, 1996.

[2] S. Bakshi and D. D. Gajski. Hardware/software partitioning and pipelining. In *Proceedings of 34th Design Automation Conference Proceedings*, 1997.

[3] R. Ernst, J. Henkel, and T. Benner. Hardware-software cosynthesis for microcontrollers. In *IEEE Design and Test of Computers*, pages 64–75, 1994.

[4] D. Gajski, F. Vahid, S. Narayan, and J. Gong. *Specification and Design of Embedded Systems*. Prentice Hall, Inc, Englewood Cliffs, New Jersey 07632, 1994.

[5] J. Gong, D. Gajski, and S. Narayan. Software estimation from executable specifications. In *The Journal of Computer and Software Engineering*, 1994.

[6] R. Gupta and G. D. Micheli. Hardware-software cosynthesis for digital systems. *IEEE Design and Test of Computers*, 10(3):29–41, 1993.

[7] T. Hu. Parallel sequencing and assembly line problems. In *Operations Research*, pages 841–848, 1961.

[8] C.-T. Hwang, Y.-C. Hsu, and Y.-L. Lin. PLS: A scheduler for pipeline synthesis. *IEEE Transactions on Computer Aided Design*, 12(9):1279–1286, Sept. 1993.

[9] K. S. Hwang, A. E. Casavant, C.-T. Chang, and M. A. d'Abreu. Scheduling and hardware sharing in pipelined data paths. In *Proceedings of the IEEE International Conference on Computer Aided Design*, pages 24–27, 1989.

[10] A. Kalavade. *System-Level Codesign Of Mixed Hardware-Software Systems*. PhD thesis, University of California, Berkeley, 1995.

[11] E. Lee and D. Messerschmitt. Synchronous data flow. *IEEE Transactions on Computers*, 75(9):1235–1245, Sept. 1987.

[12] N. Park and A. C. Parker. Sehwa: A software package for synthesis of pipelines from behavioral specifications. *IEEE Transactions on Computer Aided Design*, 7(3):356–370, Mar. 1988.

[13] A. B. Thordarson. Comparison of manual and automatic behavioral synthesis on MPEG-algorithm. Master's thesis, University of California, Irvine, 1995.

[14] F. Vahid and D. Gajski. Specification partitioning for system design. In *Proceedings of the 29th Design Automation Conference*, 1992.

Session 6

HW/SW Specification and Debugging

Co-Emulation and Debugging of HW/SW-Systems

Gernot Koch[1], Udo Kebschull[1], Wolfgang Rosenstiel[2]

[1]Forschungszentrum Informatik (FZI), Haid-und-Neu-Straße 10-14, D 76131 Karlsruhe, Germany
[2]FZI and University of Tübingen, Sand 13, D 72076 Tübingen, Germany

Abstract

In this paper we present a method that allows to observe and control the emulation of communicating systems consisting of hardware and software parts. The approach provides the ability to set breakpoints in either technology without breaking the protocol or loosing data. Systems may contain several software and hardware processes. Each process runs in a separate debugger while the presented approach provides the necessary synchronization between the processes that precludes loss of data at the communication interfaces due to breakpoints. Therefore it allows to continue the operation after a breakpoint which is essential for interactive debugging.

1. Introduction

Codesign of embedded systems is becoming more and more popular. While many research groups are working on new codesign methods, much less effort is spent on co-validation and co-debugging of such systems. In practice, mainly two technologies are used for co-validation, simulation based validation and emulation based validation.

Simulation based validation maximises the observability of a design, but offers a rather poor performance. The classical approach works with simulation models of processors for the software part and thus allows for a cycle true simulation of the whole system. This simulation can include other technologies than merely hardware and software [1]. Recent and more efficient approaches run the software on real processors and map the communication onto interprocess communication via the PLI of common hardware simulators. Since the simulator does not need to simulate the processors with this approach, the simulation is sped up significantly. [2] presents such an approach for C-VHDL co-simulation. Usually the performance of these approaches is up to a few hundred cycles per second, depending on the complexity of the hardware description.

Emulation based validation offers much less observability, but a much higher performance. Emulated systems can even run fast enough to be used within the real environment of a system as a prototype. Since the software parts run on their target processors for which development environments and debuggers exist, the observability issue concerns rather the hardware than the software. The Quickturn HDL-ICE [3] enables emulation based co-validation with observability of the hardware and backannotation of values to the RTL-description. [4] presented a co-emulation system for a fixed target architecture. It allows to set breakpoints by specifying patterns of the address bus. However, it does not allow to set functional breakpoints in the HDL-description of the hardware. Also it offers only very limited observability. The designer can evaluate memory contents, but not internal signals in the hardware. It is possible to continue operation after a breakpoint because the whole system clock is stopped at a breakpoint. However, in general this is impossible. Usually, processors and hardware run on different clocks and it is not possible to stop the clock of a processor. If the system includes DRAM, stopping the clock is also not possible since the memory cannot be refreshed then.

Since the hardware entry level is moving up to the behavioral level, breakpoints and observability on the gate level or on the RT-level are no longer sufficient. In [5] we have presented an approach which allows to control and to observe an emulated hardware by using the behavioral description. In [6] we have shown how we can set breakpoints in the behavioral description, and how we can use them for interactive debugging.

In this paper we present an approach based on this previous work, which allows for interactive debugging of co-emulated systems. In the next chapter we define our notion of communication. In chapter 3, we present how we can transform a design transparently into a debuggable design and in chapter 4 we show some examples to which we have applied our approach.

2. Communication Issues

This work concentrates on systems containing multiple processes. These processes may be software processes, running on a microprocessor, or hardware processes, designed as ASICs. All processes are synchronous processes and each of them is running on its own clock. Thus, all the processes run asynchronously. This is a typical constellation for embedded systems.

2.1. Problem definition

If an application with communicating processes runs on an emulator and one of the processes is stopped at a breakpoint while the other process isn't, data may be lost or the protocol may fail. Both cases could make it impossible to continue the operation after the breakpoint. We want to prohibit data loss or protocol failure due to one communication partner being stopped in a breakpoint, allowing to continue the operation after a breakpoint. This will of course not always be possible, but there is a reasonable number of applications and communication schemes where something can be done, as the remainder of this paper shows.

2.2. Protocol classification

Before we can go further into our solution, we need to review the communication protocols used in embedded systems. In general, we can distinguish three different, largely independent criteria. Signal synchronisation describes, how the IO-signals between the communicating partners interact during communication. It can be delay insensitive, e.g. a handshake protocol. It can also be delay sensitive, as any synchronous communication between a processor and a co-processor via the processor bus interface is. The processor just writes or reads according to its timing specifications and expects, that the co-processor can react fast enough.

Another criterion is the data synchronisation. Data synchronisation provides additional information about the data that is transferred, e.g. it tells the partner process, that data is waiting to be read or that it is ready to receive data. It may also be used to negotiate the direction of data transfer for bidirectional channels. Possible are external data synchronisation, internal synchronisation and no synchronisation. External synchronisation uses separate signals for synchronisation, while internal synchronisation uses the same signals as the data transmission. Internal synchronisation uses special characters or bitpatterns, which have to be escaped to be distinguishable from data.

The last criterion describes which type of communication is actually used. This can be RegisterIO, Interrupt, DMA, etc.Figure 1 shows the possibilities and some typical examples.

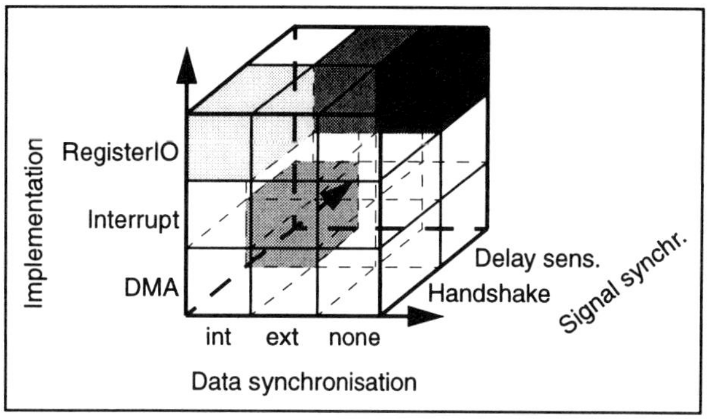

Figure 1. Protocol classification

In this figure, the -shaded block shows a bidirectional communication via a standard parallel interface. The signal synchronisation is implemented as handshake, while the data synchronisation and negotiation of the transfer direction is done as internal synchronisation. Typically, the parallel interface works as RegisterIO, since at both ends, simple registers read or write the data. The -shaded block shows a DMA. Data synchronisation is done via memory address which is not different from the data addresses, thus it is an internal synchronisation. Reading from or writing to the RAM is usually delay sensitive. The figure shows only the data communication, not the negotiation with the DMA controller. That part is usually a handshake protocol. Accordingly, shows a typical processor - co-processor communication, while is typical for a pure hardware communication. Often, the fact that data is written or read is used as data synchronisation, making explicit data synchronisation unnecessary.

2.3. Specifying communication protocols

The ability to debug communicating processes is largely depending on the specification style of communication protocols. There are basically two ways to specify protocols at the behavioral level. The protocol can be embedded in the algorithm and it can be encapsulated in a component. Figure 2 shows an example for both alternatives together with the resulting parts of the controller state machine.

Part a) shows how communication can be embedded into the algorithmic specification. The protocol in this example is a parallel interface protocol. The timing relative to the clock and the interaction of the signals is explicitly described. This has the advantage that the resulting design is quite small, but on the other hand the protocol cannot be easily exchanged and the design is only capable of receiving data, when the controller is waiting in the corresponding states. There is no computation going on in

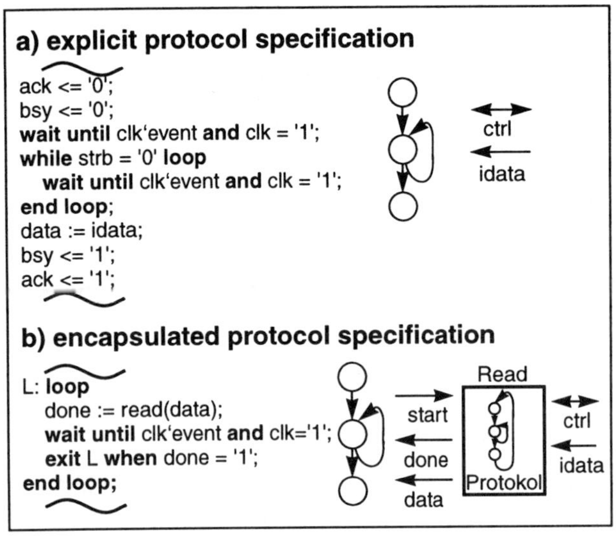

Figure 2. Protocol specification

parallel to the communication (rendevous schema). Both communicating processes meet at that time. Thus, only protocols with a handshake signal synchronisation can be implemented as shown in part a).

Part b) shows how communication protocols can be encapsulated and hidden from the algorithm. The data path controller communicates via a standard procedure with an interface component. The component is triggered with a start signal and answers back with a done signal and, if available, with data. The value of the done signal determines whether or not the data is valid. A start trigger like the start signal here is necessary for all components with an internal controller. In the HDL code this is simply represented as a procedure call.

Encapsulating the communication protocol has three advantages. First, the protocol can easily be exchanged by another protocol by simply calling another procedure, without messing up the whole algorithm. It could even be exchanged by replacing the interface component in the data path after high level synthesis is done, without changing the algorithm at all. As we will show in the next chapter, this is exactly the feature that we use to create a debuggable circuit.

Second, the interface is always ready to communicate, no matter what the data path is doing at that time. So communication and data path operation is somewhat decoupled and running in parallel. This can speed up the whole operation. Third, the designer of the algorithm does not have to be a communication expert. The interface components are pre-defined and placed into a library. There is no need to re-write them each time from scratch. As a little penalty, the resulting design size will increase somewhat. The data path controller does not change a lot (depending on the protocol of course), while the interface component needs an internal controller in addition.

3. Creating a Design-for-Debugging

3.1. Synthesis -g

As to produce a debuggable version of a design, we employ a special synthesis as shown in earlier publications. This synthesis instruments the design in a way that allows to control and observe its operation. As figure 3

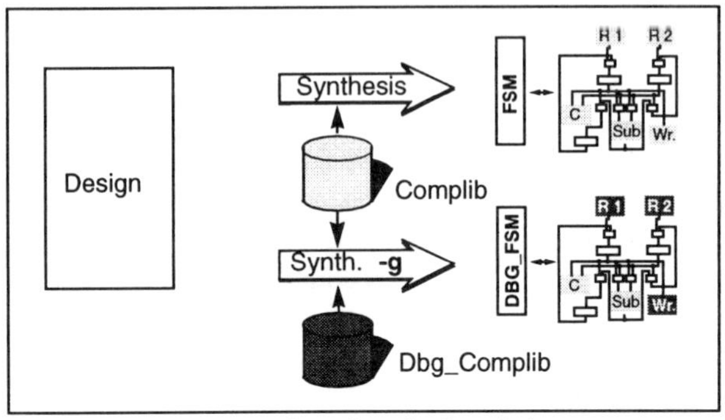

Figure 3. Synthesis for debug

shows, there is an additional component library with debug models of all sequential components like pipelined components or components with internal controllers, e.g. interface components. During normal synthesis, all components are taken from the standard library, while during synthesis for debug, called synthesis -g in the figure, all sequential components are taken from the debug library where the debug models reside. This happens automatically and is not visible to the designer.

There are two types of debug models for sequential components. The ones for purely internal components, which do only interact with the data path, simply do not change their internal state in a breakpoint. The debug models for interface components show a more sophisticated behavior. On one hand, their interface to the data path remains constant during a breakpoint, while on the other hand the communication protocol with the outside world continues to work as to fulfil the protocol.

The idea is that for each sequential component in the component library the designer also provides a debug model of the component in the debug library. In most cases doing this is trivial. However, it is not trivial for interface components

3.2. Debug models for interface components

Unfortunately we cannot provide a general rule how to derive a debug model from an arbitrary interface component. But we can provide a rule for a certain type of interface component and we can show that all common communication schemes can be implemented with components of that type. A typical protocol for reading from a standard parallel port as shown in figure 4 may serve as an illustration. It uses external data synchronisation via the

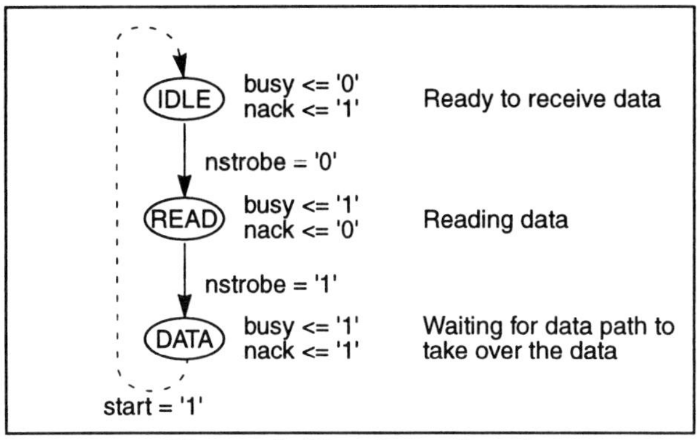

Figure 4. Simple protocol

busy signal. If busy is '1', the partner cannot send data because the component is not ready to receive it. The rest of the protocol is a 4-way handshake involving the nstrobe and nack signals. The behavior in a breakpoint is that the states IDLE and READ are processed as in normal operation, allowing the component to receive data even in a breakpoint, if it is ready to receive. In the state DATA, the transition triggered by "start = '1'" is delayed until the next active clock edge after the breakpoint. Since the value "busy"-signal is constant '1' in that state, the sender cannot send any data and has to wait until the end of the breakpoint. Thus no data is overridden or lost.

We can identify three important elements in that example. First, there is a data synchronisation through the "busy"-signal. This allows the component to prevent the sender from sending data that cannot be received during a breakpoint. Second, there is a state, where the component is waiting for the data path to act via the "start"-signal. In that state, the communication partner is held back by the data synchronisation. Third, there is a set of states, in which the component runs through the protocol. In these states, requests from the data path ("start = '1'") are answered with "done = '0'", meaning that there is no data waiting. The general cases for sending and receiving data are depicted in figure 5. In both directions there is a state, where the component is waiting for the data path to give the start trigger. In this state, data synchronisation prevents the partner process from becoming active at the interface. Only in this state, state transitions are triggered by the data path.

A component of this type can easily be transformed into a debug model by performing two steps. Step 1 delays the state transition from states like EMPTY or HAVE_DATA, which are triggered by the data path, until the next active clock edge after a breakpoint. In a breakpoint, the component can work on the protocol until it runs into a state like EMPTY or HAVE_DATA and then it stops until the operation is continued. During the stop, data synchronisation ensures, that nothing gets lost at the interface.

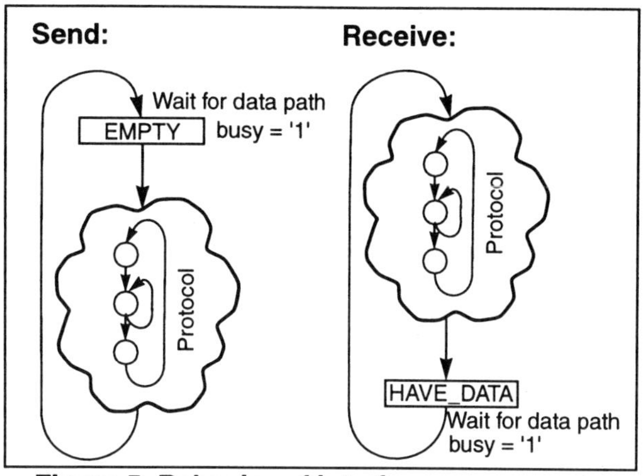

Figure 5. Behavior of interface components

Step 2 adds some hardware which holds all outputs to the data path constant during a breakpoint. This cannot be guaranteed without additional hardware, because of the possible operation that is going on during a breakpoint in the component.

Bidirectional interface components can easily be implemented by combining the state machines for sending and receiving as two independent state machines in one component. These components still follow the same rules and are transformed in the same way.

The protocol part of the state machines can implement any type of protocol. It is easily possible to implement interrupts or DMA protocols. In case of an interrupt protocol the state machine would first set an interrupt signal and then wait for the interrupt procedure to handle the data transfer.

If a DMA is implemented, the component would only handle the protocol with the DMA controller of the processor, not the data communication itself. This would be implemented via a multiport RAM or an addressable register file.

3.3. Limitations

It is obvious that there are limitations to the presented approach. In general, it cannot be applied to two types of systems. First, if the environment of the design imposes hard realtime constraints on the design, which cannot be softened for the prototype. E.g. an ethernet controller which has to listen at the bus with a certain speed cannot be interrupted at a breakpoint and still continue to work after a while. Second, if the design communicates with external devices without any data synchronisation and without a handshake protocol. In these cases, the external device cannot be manipulated by a special synthesis for debugging and thus, the presented method is not applicable. However, if two processes communicate with a protocol that uses neither data synchronisation nor handshake, but both processes are part of the synthesized design, the protocol can be transparently exchanged by a suitable protocol during synthesis as to make the prototype debuggable.

If the approach is not applicable, then there are two possible debugging strategies. One can operate with a logic analyser as it is done today, and one can run until a breakpoint occurred, evaluate the state of the system and restart from the beginning with the same parameters, but stop at another breakpoint.

4. Results

We have applied the approach to some example systems. For hardware synthesis, we have used the CADDY High Level Synthesis tool [7][8][9], which was developed at FZI. The mechanism for transparent component exchange for debugging purposes has been implemented as a "-g" switch in this tool.

The first example is a circuit that computes a 2-dimensional DCT. Data is transferred from and to a software process via DMA accesses. Figure 6 shows the structure of the example. It uses two DMA channels of the PPC403GA

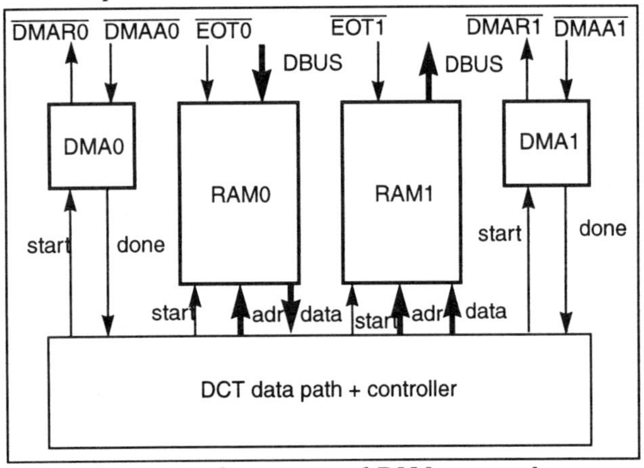

Figure 6. Structure of DMA example

[10] embedded processor, one for receiving the source data and the other for sending the transformed data. A DMA component which interacts with the DMA controller of the processor, and an internal RAM which holds the data are associated with each channel.

Figure 7 shows an excerpt from the behavioral VHDL code of the design. The first inner loop initiates the DMA request and waits until this has been acknowledged by the component. The second inner loop waits until the DMA has been finished and the data is available. Then the DCT Algorithm can start to work on the data. Writing the result back to the processor is done analogously. Both, the DMA[0,1]()-procedures and the sync[0,1]()-procedure are mapped to the DMA[0,1]-component.

The behavior of the debug model of the DMA0 component is shown in figure 8. In the IDLE state, the component waits for the data path to initiate the DMA request via a call of the procedure DMA0(). It then works through the handshake protocol with the DMA controller and ends up in the SYNC state. Now the whole DMA is finished, the data is transferred and the component waits for the data path to call the sync0()-procedure and to begin working on the data. Since IDLE and SYNC are similar states to the state "EMPTY" described in chapter 3.2, the two transitions from these states, which are triggered by the start-signal from the data path, are delayed until after the breakpoint, if the circuit is stopped in a breakpoint. The DMA1 component is treated exactly in the same way.

```
loop
  while DMA0() = '0' loop
    wait until clk'event and clk = '1';
  end loop;
  while sync0() = '0' loop
    wait until clk'event and clk = '1';
  end loop;

  -- compute the DCT

  while DMA1() = '0' loop
    wait until clk'event and clk = '1';
  end loop;
  while sync1() = '0' loop
    wait until clk'event and clk = '1';
  end loop;
end loop;
```

Figure 7. Behavioral code

The version of the design that uses an interrupt to trigger the data transfer has exactly the same structure. The only difference is in the state machine of the DMA[0,1]() component. So, from a hardware point of view, an interrupt protocol doesn't infer any further problems.

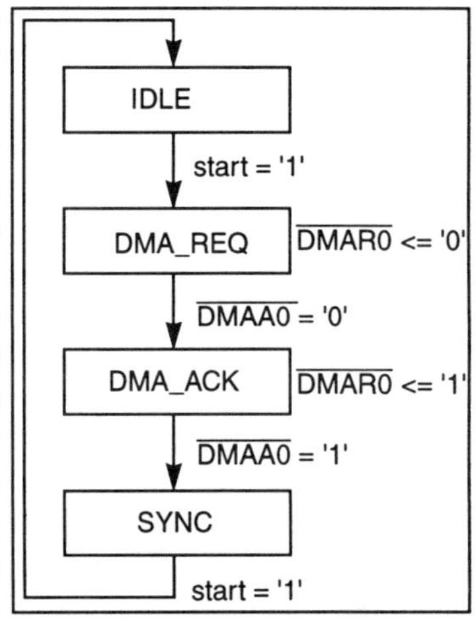

Figure 8. DMA0 component

Two other protocols have been implemented and tested. One is a bidirectional RegisterIO communication with the Hyperstone [11] processor, the other is a communication via a standard parallel interface. The RegisterIO protocol uses a status register for data synchronisation and two 32bit registers for data transfer, one for each direction. Each direction of the communication is guarded by a separate state machine, which handles the data synchronisation. Both state machines are of the type that was required in chapter 3, thus transformation from the component to the debug model of the component was done in the way that was described there.

The parallel interface protocol was implemented for both directions, each direction in a separate component. On the software side, the communication is associated with a timeout. This timeout is implemented in the operating system. When the hardware is stopped due to a breakpoint and the software wants to transfer data, then the timeout will occur. To allow debugging, the software must catch the timeout and retry to initiate the transfer in a loop. A similar protocol encapsulation in software as it is proposed for hardware in this paper allows to do that in a simple way.

We can run these designs in our debugger and do single stepping, set breakpoints and evaluate the contents of the data path registers without losing any data at the communication channel or causing a failure of the protocol.

In table 1, the area values of the different designs, which use the different protocols, are listed. All values are given in CLBs for the Xilinx XC4000 series.

Circuit	Orig.	Debug	Protocol
GCD	60	84	Hyperstone RegIO
SIRDG	99	146	Sparc10 parallel Port
DCT	387	430	PPC403GA DMA

Table 1. Area values

• **Orig.** refers to the original circuit without debug overhead.

• **Debug** includes the overhead for debugging without the possibility of data dependent breakpoints.

• **Protocol** denotes the communication protocol that was used to communicate with a software process.

Table 2 shows the delay that is added by the additional logic for debugging.

Circuit	Orig.	Debug
GCD	5.7 Mhz	4.7 Mhz
SIRDG	7.7 Mhz	4.8 Mhz
DCT	3.5 Mhz	3.5 Mhz

Table 2. Clock frequencies

The values are obtained by the Xilinx tool 'xdelay'. This tool provides a quite pessimistic estimation. All designs run at a significantly higher clock speed on our Weaver [12] board. Nevertheless, it shows the relation between the different implementations. Our approach mainly adds controller delay. Therefore we do not add any delay to the DCT, which is mainly determined by the combinational multiplier. Thus, the debug version can be clocked as fast as the original version.

5. Summary

In this paper we have presented an approach for a combined debugging of systems containing hardware and software parts. The approach is applicable to systems which use communication protocols with data synchronisation, and to systems which use handshake protocols. Systems using neither type of protocol are debuggable only, if it is possible to exchange the protocol via protocol encapsulation, when the prototype is built. We have applied the approach to some example systems to demonstrate its feasibility.

6. References

[1] Y. Tanurhan, S. Schmerler, K. Mueller-Glaser, *A Backplane Approach for Cosimulation in High-Level System Specification Environments*, European Design Automation Conference EURODAC'95, Brighton, 1995

[2] C.A. Valderrama, F. Naçabal, P. Paulin, A. A. Jerraya, *Automatic Generation of Interfaces for Distributed C-VHDL Cosimulation of Embedded Systems: an Industrial Experience*, Proceedings of 7th international Workshop on Rapid System Prototyping, Thessaloniki, 1996

[3] *ASIC-Emulation auf RTL-Level*, Markt&Technik - Wochenzeitung für Elektronik Nr. 42, 1994

[4] T. Benner, R. Ernst, I. Könenkamp, P. Schüler, H. Schaub, *A Prototyping System for Verification and Evaluation in Hardware-Software Cosynthesis,* Proceedings of 6th international Workshop on Rapid System Prototyping, Chapel Hill, 1995

[5] G. Koch, U. Kebschull, W. Rosenstiel, *Debugging of Behavioral VHDL Specifications by Source Level Emulation,* European Design Automation Conference EURODAC'95, Brighton, 1995

[6] G. Koch, U. Kebschull, W. Rosenstiel, *Breakpoints and Breakpoint Detection in Source Level Emulation*, International Symposium on System Synthesis, La Jolla, 1996

[7] R. Camposano, W. Rosenstiel, *Synthesizing Circuits from Behavioral Descriptions*, IEEE Transactions on CAD, Vol. 8, 2-1989

[8] P. Gutberlet, J. Müller, H. Krämer, W. Rosenstiel, *Automatic Module Allocation in High Level Synthesis,* European Design Automation Conference EURODAC'92, Hamburg, 1992

[9] P. Gutberlet, W. Rosenstiel, *Scheduling Between Basic Blocks in the CADDY Synthesis System,* European Conference on Design Automation EDAC'92, Brussels, 1992

[10] IBM, *PPC403GA Embedded Controller - User's Manual,* 1995

[11] Hyperstone electronics, *Hyperstone E1 32-Bit-Microprocessor User's Manual*, 1990

[12] G. Koch, U. Kebschull, W. Rosenstiel, *A Prototyping Architecture for Hardware/Software Codesign in the COBRA Project,* Proceedings of 3rd international Workshop on Hardware/Software Codesign Codes/CASHE'94, Grenoble 1994

Synthesising Controllers from Real-Time Specifications

Henning Dierks*
University of Oldenburg
Department of Computer Science
P.O.Box 2503, D-26111 Oldenburg, Germany
dierks@informatik.uni-oldenburg.de

Abstract

We present an algorithm for synthesising controllers specified in a subset of the interval temporal logic Duration Calculus [13]. The synthesised controllers are given as PLC-Automata [5] and these can be implemented directly on PLCs [5]. We demonstrate the behaviour of the algorithm by an example.

1. Introduction

Provable correct software for embedded real-time systems is a main topic of recent research (see e.g. [9, 7]). All approaches try to develop correct systems either by refinement, model-checking, or by synthesis from specifications. In the case of model-checking one has to handle the trade-off between the expressiveness of the logic and runtime of the model-checking algorithm. Furthermore, the specification logic has to be decidable. One successful approach to real-time systems are Timed Automata [2]. Although quite expressive real-time logics exist [1] for Timed Automata and even tools which implement model-checking algorithms for Timed Automata [4], it is not clear how to implement Timed Automata on real hardware.

Another approach which falls into the refinement category is the ProCoS-method [6] that starts with specifications in the interval temporal logic Duration Calculus [13] (DC for short) and refines through several steps down to Transputer hardware. From the scientific point of view this project was successful, but it is hardly applicable in industry because there is no tool support for users of this method yet. The succeeding UniForM-project [8] tries to overcome this disadvantage by adopting the transformational ideas and methodologies of the ProCoS-project and supplying a tool, the UniForM Workbench, that supports the users of the UniForM-approach.

One main difference between ProCoS and UniForM is the hardware the approaches are aiming at. While ProCoS aims at Transputers and an OCCAM-like language, UniForM develops source code for PLCs (Programmable Logic Controllers), a widely used hardware in industry to control real-time systems like railway crossings or production lines. These machines behave in cycles that can be split into three parts: the input ports are polled, the new outputs are computed from the read input values, and finally the new output values are written to the output ports. The software designer only has to change the computing part while the other parts are managed by the operation system of the PLC. Depending on the computing part there is an upper time bound for a PLC's cycle. It turned out that "PLC-Automata", a certain description technique for PLC-programs with a DC-semantics [5], are very useful as an intermediate language because it bridges the gap between the specification language DC and PLCs.

Thus the question arose how to come from arbitrary DC specifications to PLC-Automata. In the ProCoS-project a certain subset of DC formulas, so called "Implementables" [11], turned out to be useful as a stepping stone for specifying distributed real-time systems. In [11] a fully developed theory can be found how Implementables can be obtained from general DC formulas. Hence, a general method to obtain PLC-Automata from Implementables would be a powerful means to design correct systems. In this paper we present an algorithm that synthesises a PLC-Automaton from an arbitrary set of Implementables provided that this set is consistent. That means it constrains the output from the input and not vice versa, and it does not require instantaneous reaction of the system. Hence, we get a synthesis procedure that produces source code for existing machines that is provably correct with respect to the DC-specification. So our method uses transformational refinement to obtain Implementables from DC-specifications and uses afterwards the synthesis algorithm given in this paper instead of further

*This research was supported by the German Ministry for Education and Research (BMBF) as part of the project UniForM under grant No. FKZ 01 IS 521 B3

refinement steps via several languages as in the ProCoS-project.

2. Duration Calculus

For the sake of completeness we recall the DC but in this paper we use only a subset of DC called "Implementables" which are explained informally in Section 3. DC is a real-time interval temporal logic extending earlier work on discrete interval temporal logic of [10].

A formal description of a real-time system using DC starts by choosing a number of time-dependent state variables or observables *obs* of a certain type. An interpretation $\mathcal{I}$ assigns to each state variable a function $obs_{\mathcal{I}} : \mathsf{Time} \longrightarrow D$ where Time is the time domain, here the non-negative reals, and D is the value domain of *obs*. The function $obs_{\mathcal{I}}$ is *finitely variable*, i.e. it has only finitely many points of discontinuity in each finite interval and is constant between them.

State assertions P are obtained by applying propositional connectives to elementary assertions of the form $obs = v$ for a $v \in D$. We often abbreviate $obs = v$ with v if *obs* is clear. In the case of a disjunction of elementary assertions $obs = v_1 \vee \ldots \vee obs = v_n$ we write $obs \in \{v_1, \ldots, v_n\}$ (or $\{v_1, \ldots, v_n\}$ for short). For a given interpretation $\mathcal{I}$ state assertions denote functions $P_{\mathcal{I}} : \mathsf{Time} \longrightarrow Bool$.

Duration formulae F are evaluated in a given interpretation $\mathcal{I}$ and a given time interval $[b, e] \subseteq \mathsf{Time}$. The basic syntax of duration formulae is as follows (according to their priority):

- **Duration:** $\int P = k$ expresses that the *duration* of the state assertion P in $[b, e]$ is k. Semantically, duration is the integral $\int_b^e P_{\mathcal{I}}(t)dt$.
- **Chop:** The composite duration formula $F_1; F_2$ (read as F_1 *chop* F_2) holds in $[b, e]$, if this interval can be divided into an initial subinterval $[b, m]$ where F_1 holds and a final subinterval $[m, e]$ where F_2 holds.
- **Connectives:** Duration formulae are closed under propositional connectives like $\neg$, $\wedge$, and $\vee$.

Besides this basic syntax various abbreviations are used:

length:	ℓ	$\stackrel{\text{df}}{=}$	$\int true$
point interval:	$\lceil\rceil$	$\stackrel{\text{df}}{=}$	$\ell = 0$
everywhere:	$\lceil P \rceil$	$\stackrel{\text{df}}{=}$	$(\int P = \ell \wedge \ell > 0)$
somewhere:	$\Diamond F$	$\stackrel{\text{df}}{=}$	$true; F; true$
always:	$\Box F$	$\stackrel{\text{df}}{=}$	$\neg\Diamond\neg F$

A duration formula F *holds* in an interpretation $\mathcal{I}$ if F evaluates to true in $\mathcal{I}$ and every interval of the form $[0, t]$ with $t \in \mathsf{Time}$. The following so-called *standard forms* [11] are useful to describe dynamic behaviour:

- *followed-by*: $F \longrightarrow \lceil P \rceil \stackrel{\text{df}}{=} \Box\neg(F; \lceil \neg P \rceil)$
- *timed leads-to*: $F \stackrel{t}{\longrightarrow} \lceil P \rceil \stackrel{\text{df}}{=} (F \wedge \ell = t) \longrightarrow \lceil P \rceil$
- *timed up-to*: $F \stackrel{\leq t}{\longrightarrow} \lceil P \rceil \stackrel{\text{df}}{=} (F \wedge \ell \leq t) \longrightarrow \lceil P \rceil$

As before we have $t \in \mathsf{Time}$. Intuitively, $F \longrightarrow \lceil P \rceil$ expresses the fact that whenever a pattern given by a formula F is observed, then it will be "followed by" an interval in which P holds. In the "leads-to" form this pattern is required to have a length t, and in the "up-to" form the pattern is bounded by a length up to t. Note that the "leads-to" does not simply say that whenever F holds then t time units later $\lceil P \rceil$ holds; rather, a stability of F for t time units is required before we can be certain that $\lceil P \rceil$ holds. The "up-to" form is mainly used to specify certain stability conditions. For example $\lceil \neg\pi \rceil ; \lceil \pi \rceil \stackrel{\leq t}{\longrightarrow} \lceil \pi \rceil$ is an expression that is true iff π is stable for at least t seconds whenever π becomes true. Note that the standard forms have the weakest priority.

3. Implementables

One result of the ProCoS-project is that a certain subset of DC-formulas is useful to specify the behaviour of controllers. These formulas are called "Implementables". They speak about an input observable with domain Σ and an output observable with a finite domain Π. We explain the Implementables using the following example:

Example 3.1 *Consider a sensor for a track segment with output* $\{\mathtt{n}, \mathtt{t}\}$ *where* n *stands for "no train detected" and* t *stands for "train detected". We want to build a system that recognises trains and raises an error, whenever a train remains in the track segment for longer than 10 seconds. The output of the system should be* $\{\mathtt{N}, \mathtt{T}, \mathtt{E}\}$ *standing for "No train on track", "Train on track", and "Error".*

The following types of formulas are "Implementables":
Initialisation: $\lceil\rceil \vee \lceil \pi_0 \rceil ; \mathsf{true}$ with $\pi_0 \in \Pi$ says that the output is initially π_0. We want our example system to start in state N. Thus, we specify $\lceil\rceil \vee \lceil \mathtt{N} \rceil ; \mathsf{true}$.
Sequencing: $\lceil \pi \rceil \longrightarrow \lceil \pi \vee \Gamma \rceil$ with $\pi \in \Pi$ and $\Gamma \subseteq \Pi$ postulates that if the output is π it can only change to an output in Γ. In the example we want that after N only T is possible: $\lceil \mathtt{N} \rceil \longrightarrow \lceil \mathtt{N} \vee \{\mathtt{T}\} \rceil$. Furthermore we wish that the E-state holds forever ($\lceil \mathtt{E} \rceil \longrightarrow \lceil \mathtt{E} \vee \emptyset \rceil$). Finally we specify that all states are possible when leaving T: $\lceil \mathtt{T} \rceil \longrightarrow \lceil \mathtt{T} \vee \{\mathtt{N}, \mathtt{E}\} \rceil$.
Unbounded Stability: $\lceil \neg\pi \rceil ; \lceil \pi \wedge \phi \rceil \longrightarrow \lceil \pi \vee \Phi \rceil$ with $\pi \in \Pi$, $\emptyset \neq \phi \subseteq \Sigma$, and $\Phi \subseteq \Pi$ says that the output can only change from π to an output in Φ if the only inputs read since the change of the output to π are in ϕ. In the example we can use this to specify that N holds as long as n is read ($\lceil \neg\mathtt{N} \rceil ; \lceil \mathtt{N} \wedge \{\mathtt{n}\} \rceil \longrightarrow \lceil \mathtt{N} \vee \emptyset \rceil$) and to specify that E is not possible after T if only n is read: $\lceil \neg\mathtt{T} \rceil ; \lceil \mathtt{T} \wedge \{\mathtt{n}\} \rceil \longrightarrow \lceil \mathtt{T} \vee \{\mathtt{N}\} \rceil$. Furthermore we want

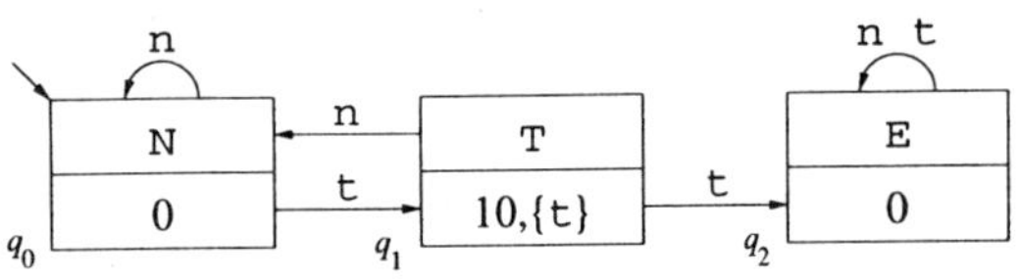

Figure 1. Example 3.1 as PLC-Automaton.

that in state $\mathtt{T}$ only changes to $\mathtt{E}$ are possible if only $\mathtt{t}$ is read. $\lceil\neg\mathtt{T}\rceil\,;\lceil\mathtt{T}\wedge\{\mathtt{t}\}\rceil\longrightarrow\lceil\mathtt{T}\vee\{\mathtt{E}\}\rceil$.

Bounded Stability: $\lceil\neg\pi\rceil\,;\lceil\pi\wedge\psi\rceil\xrightarrow{\leq t}\lceil\pi\vee\Psi\rceil$ with $\pi\in\Pi$, $t>0$, $\emptyset\neq\psi\subseteq\Sigma$, and $\Psi\subseteq\Pi$. This restricts changes from π like the unbounded stability but just for the first t seconds after the change to π. This Implementable can be used to require that $\mathtt{T}$ holds at least 10 seconds if only $\mathtt{t}$ is signalled by the sensor: $\lceil\neg\mathtt{T}\rceil\,;\lceil\mathtt{T}\wedge\{\mathtt{t}\}\rceil\xrightarrow{\leq 10}\lceil\mathtt{T}\vee\emptyset\rceil$.

Synchronisation: $\lceil\pi\wedge\varphi\rceil\xrightarrow{s}\lceil\neg\pi\rceil$ with $\pi\in\Pi$, $s>0$, and $\emptyset\neq\varphi\subseteq\Sigma$ says that if the input is in φ for s seconds while the output is π then the output changes. In the example we want our system to react after at most γ seconds for some parameter $\gamma>0$. Therefore, we specify that we leave $\mathtt{N}$ (resp. $\mathtt{T}$) when reading $\mathtt{t}$ (resp. $\mathtt{n}$): $\lceil\mathtt{N}\wedge\{\mathtt{t}\}\rceil\xrightarrow{\gamma}\lceil\neg\mathtt{N}\rceil$ and $\lceil\mathtt{T}\wedge\{\mathtt{n}\}\rceil\xrightarrow{\gamma}\lceil\neg\mathtt{T}\rceil$. Furthermore, we want to leave $\mathtt{T}$ after at most $10+\gamma$ seconds: $\lceil\mathtt{T}\wedge\{\mathtt{n},\mathtt{t}\}\rceil\xrightarrow{10+\gamma}\lceil\neg\mathtt{T}\rceil$.

4. PLC-Automata

In the UniForM-project [8] we made the experience that automaton-like pictures can serve as a common basis for computer scientists and engineers. The latter have a intuitive understanding of these pictures. Therefore, we formalised the notion of "PLC-Automaton" and defined a formal semantics for it in DC (App. A) that is consistent to the translation of PLC-Automata into PLC-source-code. But PLC-Automata are not especially tailored to PLCs. In fact, PLC-Automata are an abstract representation of a machine that periodically polls the input and has the possibility of measuring time. In [5] a compilation function to PLC source code is given.

Figure 1 gives an example of a PLC-Automaton. It shows an implementation of Example 3.1 with three states ($\{q_0,q_1,q_2\}$) and outputs $\{\mathtt{N},\mathtt{T},\mathtt{E}\}$, that reacts to inputs of the alphabet $\{\mathtt{n},\mathtt{t}\}$. Every state has two annotations in the graphical representation. The upper one denotes the output of the state, thus in state q_0 the output is $\mathtt{N}$ and in state q_2 the output is $\mathtt{E}$ denoting an "Error". The lower annotation is either 0 or a pair consisting of a real number $d>0$ and a nonempty subset S of inputs.

The operational behaviour is as follows: If the second annotation of a state q is 0, the PLC-Automaton reacts in every cycle to the inputs that are read and behaves according to the transition relation. If the second annotation of q is a pair (d,S), the PLC-Automaton checks in every cycle the input i according to these parameters. If i is not in S the automaton reacts immediately according to the transition relation. If i is in S and the current state does not hold longer than d, the input will be ignored and the automaton remains in state q. If i is in S and state q held longer than d the PLC-Automaton will react on i according to the transition relation.

The PLC-Automaton in Fig. 1 thus behaves as follows: It starts in state q_0 and remains there as long as it reads only the input $\mathtt{n}$. The first time it reads $\mathtt{t}$ it changes to state q_1. In q_1 the automaton reacts to the input $\mathtt{n}$ by changing the state back to q_0 independently of the time it stayed in state q_1. It reacts to the input $\mathtt{t}$ by changing the state to q_2 provided that q_1 holds longer than 10 seconds. If this transition takes place the automaton enters q_2 and remains there forever. Hence, we know that the automaton changes its output to $\mathtt{E}$ when $\mathtt{t}$ holds a little bit longer than 10 seconds (the cycle time has to be considered).

We formalise this graphic notation using an automaton-like structure extended by some components:

Definition 4.1 *A tuple* $\mathcal{A}=(Q,\Sigma,\delta,\pi_0,\varepsilon,S_t,S_e,\Omega,\omega)$ *is a* PLC-Automaton *if*

- Q *is a nonempty, finite set of* states,
- Σ *is a nonempty, finite set of* inputs,
- δ *is a function of type* $Q\times\Sigma\longrightarrow Q$ *(*transition function*)*,
- $\pi_0\in Q$ *is the* initial state,
- $\varepsilon>0$ *is the* upper bound *for a cycle,*
- S_t *is a function of type* $Q\longrightarrow\mathbf{R}_{\geq 0}$ *assigning to each state* π *a* delay time *how long the inputs contained in* $S_e(\pi)$ *should be ignored,*
- S_e *is a function of type* $Q\longrightarrow\mathcal{P}(\Sigma)\setminus\{\emptyset\}$ *assigning to each state a set of* delayed inputs,[1]
- Ω *is a nonempty, finite set of* outputs,
- ω *is a function of type* $Q\longrightarrow\Omega$ *(*output function*)*

and it holds for all $\pi\in Q$ *and* $a\in\Sigma$*:*

$$S_t(\pi)>0\wedge a\notin S_e(\pi)\Longrightarrow\delta(\pi,a)\neq\pi. \tag{1}$$

The components Q, Σ, δ, and q_0 are as for usual finite state automata. The additional components are needed to model PLC-behaviour and to enrich the language for dealing with real-time aspects. The ε represents the upper bound for a cycle of a PLC and enables us to model this cycle in the semantics. The delay function S_t and S_e represent the annotations of the states. In the case of $S_t(\pi)=0$ no delay time is given and the value $S_e(\pi)$ is arbitrary. If the delay time $S_t(\pi)$ is greater than 0 the set $S_e(\pi)$ denotes the set of inputs for which the delay time is valid. The restriction (1)

[1] If $S_t(\pi)=0$ the set $S_e(\pi)$ can be arbitrarily chosen. The single 0 represents this in the graphical notation (cf. Fig. 1).

is introduced to enable the application of the compilation schema given in [5].

5. The Synthesis Algorithm

In this section we present an algorithm which computes a PLC-Automaton from a set of Implementables provided that this set is consistent. We demonstrate the operations of the algorithm by Example 3.1 with the set of Implementables given in Sect. 3.

The idea of the algorithm is that we collect for each output state π the Implementables speaking about π. We sort the time bounds appearing in all bounded stabilities for π and build a cascade of internal automaton states all with output π. The first internal state of the cascade has to consider all bounded stabilities, the second one only those bounded stabilities with bounds that are not the smallest one, etc. The last internal state of the cascade has no bounded stability to consider. After building this cascade we build up two tables. The first table represents the possible transitions in an internal state depending on the read input. The second table contains the information if an input is a delayed input in an internal state. All Implementables are now used to make changes in the table in order to ensure that the cascade fulfils the requirement. Thus, the Implementables are translated into requirements for the tables resp. for the cascade. Finally, we compute for π an upper bound of the cycle time from the timed constraints.

In more detail the algorithm works as follows: Except for the initialisation all Implementables speak about one certain state π of a controller. Hence, we construct a part of a PLC-Automaton that implements the behaviour of the state π. Assume that the specification of the controller expressed in Implementables contains these formulas for π:

$$\begin{aligned} \lceil\pi\rceil &\longrightarrow \lceil\pi\vee\Gamma\rceil \\ \forall i\in I: \lceil\neg\pi\rceil\,;\lceil\pi\wedge\phi_i\rceil &\longrightarrow \lceil\pi\vee\Phi(\phi_i)\rceil \\ \forall j\in J: \lceil\neg\pi\rceil\,;\lceil\pi\wedge\psi_j\rceil &\xrightarrow{\leq t'_j} \lceil\pi\vee\Psi(\psi_j)\rceil \\ \forall k\in K: \lceil\pi\wedge\varphi_k\rceil &\xrightarrow{s_k} \lceil\neg\pi\rceil \end{aligned}$$

The sets I, J, K are finite (and possibly empty). Wlog. we assume that $\Phi(\phi_i)$ and $\Psi(\psi_k)$ are subsets of Γ. Assume that $\{t_1,\dots,t_{n(\pi)}\} = \{t'_j | j\in J\}$ with $t_1 < t_2 < \cdots < t_{n(\pi)}$. We put $t_0 \stackrel{\text{df}}{=} 0$ and $t_{n(\pi)+1} \stackrel{\text{df}}{=} \infty$. In the example we have $n(\text{N}) = n(\text{E}) = 0$ and $n(\text{T}) = 1$ with $t_1 = 10$. The part of the PLC-Automaton we construct for π is a cascade of internal states $q_{\pi,t_1},\dots,q_{\pi,t_{n(\pi)}},q_{\pi,\infty}$. The idea is that the state q_{π,t_i} represents the knowledge that π has held at least t_{i-1} seconds and at most $t_i + 2i\varepsilon$ seconds with $\varepsilon > 0$ representing the cycle time of the PLC-Automaton. Note that in the worst case it lasts one cycle to recognise that the delay time is up and another cycle to react to the input. Hence, in the worst case we pick up in each internal state of the cascade 2ε seconds.

Whenever the PLC-Automaton changes its output to π it starts in the state q_{π,t_1} and for all q_{π,t_i} with $t_i \neq \infty$ transition to a q_{π,t_j} with $j \leq i$ or $j > i+1$ are not allowed. For $q_{\pi,\infty}$ transitions to a q_{π,t_j} with $j \leq n(\pi)$ are forbidden. Hence the skeleton of this part of the PLC-Automaton looks as follows (with $n \stackrel{\text{df}}{=} n(\pi)$):

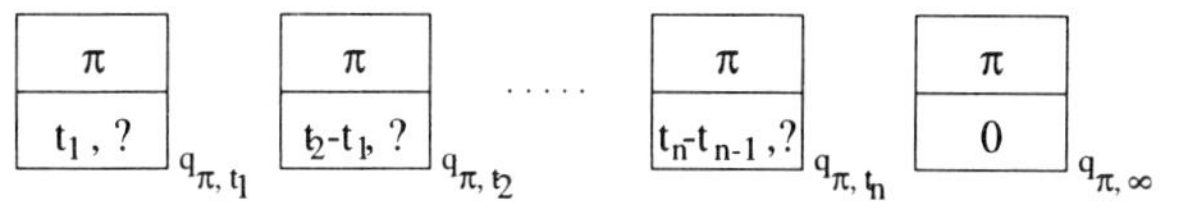

We have now to determine how the transition function works and which inputs are delayed.

Step (0): Build the following tables δ^π and S_e^π (suppose that the set of inputs Σ is $\{\kappa_1,\dots,\kappa_m\}$ and let $\Gamma_\pi \stackrel{\text{df}}{=} \Gamma\cup\{\pi\}$.):

δ^π	t_1	$\dots$	$t_{n(\pi)+1}$
κ_1	Γ_π	$\dots$	Γ_π
$\vdots$	$\vdots$	$\ddots$	$\vdots$
κ_m	Γ_π	$\dots$	Γ_π

S_e^π	t_1	$\dots$	$t_{n(\pi)}$
κ_1	true	$\dots$	true
$\vdots$	$\vdots$	$\ddots$	$\vdots$
κ_m	true	$\dots$	true

The $\delta^\pi(\kappa_l, t_k)$-entry represents the possible transitions from state q_{π,t_k} when reading the input κ_l. The $S_e^\pi(\kappa_l, t_k)$-entry contains the information whether in state q_{π,t_k} the input κ_l is delayed or not. We know that no more input is delayed after $t_{n(\pi)}$ seconds because there is no bounded stability with a larger bound. Hence,we omit a $t_{n(\pi)+1}$-column in this table. Initially we allow only transitions to Γ_π and delay all inputs. In our example we have these sequencings:

$$\lceil\text{N}\rceil \longrightarrow \lceil\text{N}\vee\{\text{T}\}\rceil \qquad \lceil\text{E}\rceil \longrightarrow \lceil\text{E}\vee\emptyset\rceil$$
$$\lceil\text{T}\rceil \longrightarrow \lceil\text{T}\vee\{\text{N},\text{E}\}\rceil$$

Hence, we get these tables:

δ^{N}	∞
n	N, T
t	N, T

δ^{T}	10	∞
n	N, T, E	N, T, E
t	N, T, E	N, T, E

δ^{E}	∞
n	E
t	E

S_e^{T}	10
n	true
t	true

The other Implementables are taken into account in the following steps:

Step (1): Replace for all $\lceil\neg\pi\rceil\,;\lceil\pi\wedge\phi_i\rceil \longrightarrow \lceil\pi\vee\Phi(\phi_i)\rceil$ in the specification and for all $\kappa_l\in\phi_i$ all entries $\delta^\pi(\kappa_l, t_j)$ in the δ^π-table by $\delta^\pi(\kappa_l, t_j)\cap(\Phi(\phi_i)\cup\{\pi\})$. This avoids transition from π to a state not in $\Phi(\phi_i)$ when reading an input in ϕ_i. In the example we have three Implementables of this kind:

$$\lceil\neg\text{N}\rceil\,;\lceil\text{N}\wedge\{\text{n}\}\rceil \longrightarrow \lceil\text{N}\vee\emptyset\rceil$$
$$\lceil\neg\text{T}\rceil\,;\lceil\text{T}\wedge\{\text{n}\}\rceil \longrightarrow \lceil\text{T}\vee\{\text{N}\}\rceil$$
$$\lceil\neg\text{T}\rceil\,;\lceil\text{T}\wedge\{\text{t}\}\rceil \longrightarrow \lceil\text{T}\vee\{\text{E}\}\rceil$$

This step yields the following δ^π-tables of Example 3.1:

δ^{N}	∞
n	N, $\not{\mathrm{T}}$
t	N, T

δ^{T}	10	∞
n	N, T, $\not{\mathrm{E}}$	N, T, $\not{\mathrm{E}}$
t	$\not{\mathrm{N}}$, T, E	$\not{\mathrm{N}}$, T, E

δ^{E}	∞
n	E
t	E

S_e^{T}	10
n	true
t	true

Step (2): Replace for all $\lceil \neg\pi \rceil ; \lceil \pi \wedge \psi_j \rceil \xrightarrow{\leq t'_j} \lceil \pi \vee \Psi(\psi_j) \rceil$ in the specification and for all $\kappa_l \in \psi_j$ all entries $\delta^\pi(\kappa_l, t_{j'})$ in the δ^π-table by $\delta^\pi(\kappa_l, t_{j'}) \cap (\Psi(\psi_j) \cup \{\pi\})$ provided that $t_{j'} \leq t'_j$. This avoids transition from π to a state not in $\Psi(\psi_j)$ when reading an input in ψ_j before t_j seconds have elapsed. In the example we have only one bounded stability:

$$\lceil \neg \mathtt{T} \rceil ; \lceil \mathtt{T} \wedge \{\mathtt{t}\} \rceil \xrightarrow{\leq 10} \lceil \mathtt{T} \vee \emptyset \rceil .$$

Thus, we get now:

δ^{N}	∞
n	N
t	N, T

δ^{T}	10	∞
n	N, T	N, T
t	T, $\not{\mathrm{E}}$	T, E

δ^{E}	∞
n	E
t	E

S_e^{T}	10
n	true
t	true

Step (3): Replace for all $\lceil \pi \wedge \varphi_k \rceil \xrightarrow{s_k} \lceil \neg\pi \rceil$ and for all $\kappa_l \in \varphi_k$ all entries $\delta^\pi(\kappa_l, t_j)$ in the δ^π-table by $\delta^\pi(\kappa_l, t_j) \setminus \{\pi\}$ provided that $s_k \leq t_j$. This avoids that the state π is held reading input κ_l in the case that too much time has already elapsed. Three synchronisations are given in the example:

$$\lceil \mathtt{N} \wedge \{\mathtt{t}\} \rceil \xrightarrow{\gamma} \lceil \neg \mathtt{N} \rceil \qquad \lceil \mathtt{T} \wedge \{\mathtt{n}\} \rceil \xrightarrow{\gamma} \lceil \neg \mathtt{T} \rceil$$

$$\lceil \mathtt{T} \wedge \{\mathtt{n}, \mathtt{t}\} \rceil \xrightarrow{10+\gamma} \lceil \neg \mathtt{T} \rceil$$

Now the δ^π-tables for the example look like this:

δ^{N}	∞
n	N
t	$\not{\mathrm{N}}$, T

δ^{T}	10	∞
n	N, $\not{\mathrm{T}}$	N, $\not{\mathrm{T}}$
t	T	$\not{\mathrm{T}}$, E

δ^{E}	∞
n	E
t	E

S_e^{T}	10
n	true
t	true

Step (4): Finally, set for all $\lceil \pi \wedge \varphi_k \rceil \xrightarrow{s_k} \lceil \neg\pi \rceil$ and for all $\kappa_l \in \varphi_k$ the $S_e^\pi(\kappa_l, t_j)$-entry in the S_e^π-table to false provided that $s_k \leq t_j$. This is the only step which modifies the S_e^π-table. It avoids that an input in φ_k is delayed after s_k seconds have elapsed. So we get:

δ^{N}	∞
n	N
t	T

δ^{T}	10	∞
n	N	N
t	T	E

δ^{E}	∞
n	E
t	E

S_e^{T}	10
n	false
t	true

Note that except for step (0) the order of the other steps is arbitrary .

Lemma 5.1 *For all π and for all κ_l, t_k the following holds:* $\pi \in \delta^\pi(\kappa_l, t_k) \iff S_e^\pi(\kappa_l, t_k) = \text{true}$

This lemma is obvious due to the construction steps (3) and (4). It guarantees that a transition from q_{π,t_i} to $q_{\pi,t_{i+1}}$ is only possible after the delay time of q_{π,t_i}. This preserves our intuition that in state q_{π,t_i} the output π held at least t_{i-1} seconds and at most $t_i + 2i\varepsilon$ seconds.

We define now $\varepsilon(\pi) \in \mathbf{R}_{>0}$, which represents the upper bound for the cycle time of the PLC-Automaton to guarantee the timing constraints for π. It is the maximal number ξ such that

$$\xi \leq \frac{s_k}{3} \tag{2}$$

holds for all $k \in K$ and for all $j \in J$ with $s_k > t_j$ it holds

$$\xi \leq \frac{s_k - t_j}{2j + 1} . \tag{3}$$

If $K = \emptyset$ we set $\varepsilon(\pi) = \infty$. Bound (2) is needed because in the worst case a PLC-Automaton needs three cycles to react to an input. In the presence of bounded stabilities we have to take into account that state q_{π,t_j} only represents the knowledge that π held at most $t_j + 2j\varepsilon$ seconds. In this case the automaton needs one more cycle to react which lead us to the inequality $t_j + (2j+1)\varepsilon \leq s_k$ in order to react in time. Hence, we get bound (3). In the example we have $\varepsilon(\mathrm{N}) = \frac{\gamma}{3}$, $\varepsilon(\mathrm{T}) = \frac{\gamma}{3}$, and $\varepsilon(\mathrm{E}) = \infty$. Note, that $\varepsilon(\pi) > 0$ always exists.

Hence, we get for each state π a pair of tables δ^π and S_e^π. We now construct a PLC-Automaton out of these tables supposing that Π is the set of output states the Implementables are speaking about.

Definition 5.2 *Given a set Π of output states of a controller and a set $SPEC$ of Implementables speaking about Π using the input observable* input : Time $\longrightarrow \Sigma$, *we call each PLC-Automaton $\mathcal{A} = (Q, \Sigma, \delta, q_0, \varepsilon, S_t, S_e, \Pi, \omega)$ a $SPEC$-Automaton if it has the following properties:*

$$\begin{array}{rcl} Q &=& \{q_{\pi,t_j} | \pi \in \Pi, 1 \leq j \leq n(\pi)+1\} \\ \varepsilon &<& \min\{\varepsilon(\pi) | \pi \in \Pi\} \\ S_t(q_{\pi,t_j}) &=& \begin{cases} 0, & j = n(\pi)+1 \\ t_j - t_{j-1}, & 1 \leq j \leq n(\pi) \end{cases} \\ S_e(q_{\pi,t_j}) &=& \{\kappa_l \in \Sigma | S_e^\pi(\kappa_l, t_j) = \text{true}\} \\ \omega(q_{\pi,t_j}) &=& \pi \end{array}$$

The transition relation δ has to fulfil the following: for all $\kappa_l \in \Sigma$ and for all $\pi \in \Pi$, $1 \leq j \leq n(\pi)+1$ and $q_{\pi',t_{j'}} = \delta(q_{\pi,t_j}, \kappa_l)$ holds

$$\begin{array}{l} \pi' \in \delta^\pi(\kappa_l, t_j) \\ \wedge\ \pi \neq \pi' \Longrightarrow j' = 1 \\ \wedge\ \pi = \pi' \Longrightarrow (j' = j+1 \vee j' = j = n(\pi)+1). \end{array}$$

If there is an initial constraint $\lceil\rceil \vee \lceil \pi_0 \rceil$; true in $SPEC$ the initial state q_0 is q_{π_0,t_1}. Otherwise q_0 is in $\{q_{\pi,t_1} | \pi \in \Pi\}$.

In the rest of this paper we will define what a consistent specification is (Def. 5.3) and show that for every consistent specification $SPEC$ a $SPEC$-Automaton exists (Theorem 5.4). Finally, Theorem 5.5 shows that every $SPEC$-Automaton satisfies $SPEC$. Figure 2 shows an automaton

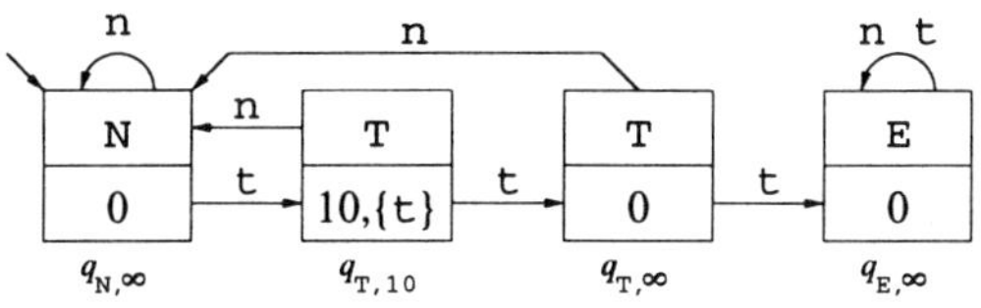

Figure 2. The result of the algorithm.

with cycle time $\varepsilon \leq \frac{\gamma}{3}$ which is obtained by the above synthesis algorithm from the Implementables of Section3. Note that this automaton differs slightly from the automaton in Fig. 1. It is not guaranteed that the algorithm produces an optimal result. But the states $q_{\pi,\infty}$ are the only possible superfluous states which are introduced by the algorithm. All other superfluous states are introduced by the specification itself when containing superfluous bounded stabilities.

Definition 5.3 *We call a specification using Implementables* consistent *if there is at most one Initialisation requirement and for every synchronisation* $\lceil \pi \wedge \varphi \rceil \xrightarrow{s} \lceil \neg\pi \rceil$ *the following holds: for any input* $i \in \varphi$ *there is a state* $\pi' \neq \pi$ *such that*

- *for every* $\lceil \pi \rceil \longrightarrow \lceil \pi \vee \Gamma \rceil$ *it holds:* $\pi' \in \Gamma$.
- *for every* $\lceil \neg\pi \rceil ; \lceil \pi \wedge \phi \rceil \longrightarrow \lceil \pi \vee \Phi(\phi) \rceil$ *it holds:* $i \in \phi \Longrightarrow \pi' \in \Phi(\phi)$.
- *for every* $\lceil \neg\pi \rceil ; \lceil \pi \wedge \psi \rceil \xrightarrow{\leq t} \lceil \pi \vee \Psi(\psi) \rceil$ *it holds:* $i \in \psi \wedge t \geq s \Longrightarrow \pi' \in \Psi(\psi)$.

Theorem 5.4 *For each consistent specification* $SPEC$ *of a controller that speaks about* Π, *with the input observable* input, *there exists a* $SPEC$*-Automaton.*

Proof: Only three things have to be checked: restriction (1), $\varepsilon > 0$ and for every π, κ_l and t_k the $\delta^\pi(\kappa_l, t_k)$-entry is not empty. The restriction (1) is fulfilled which can be seen by Lemma 5.1. The second obligation is trivial by the definition of $\varepsilon(\pi)$ which is always greater 0. Therefore the minimum of these bounds is also greater 0 because we have a finite set of output states. The last obligation can be seen as follows: Suppose $\pi \notin \delta^\pi(\kappa_l, t_k)$. Then we know that there was a synchronisation constraint $\lceil \pi \wedge \varphi \rceil \xrightarrow{s} \lceil \neg\pi \rceil$ with $\kappa_l \in \varphi$ and $s \leq t_k$. By the assumption that $SPEC$ is consistent we get a π' that is by construction in $\delta^\pi(\kappa_l, t_k)$. □

Theorem 5.5 *Given a consistent specification* $SPEC$ *of a controller with the input observable* input *and the set of output states* Π, *any* $SPEC$*-Automaton* $\mathcal{A}$ *implements* $SPEC$.

A sketch of the proof can be found in App. B.

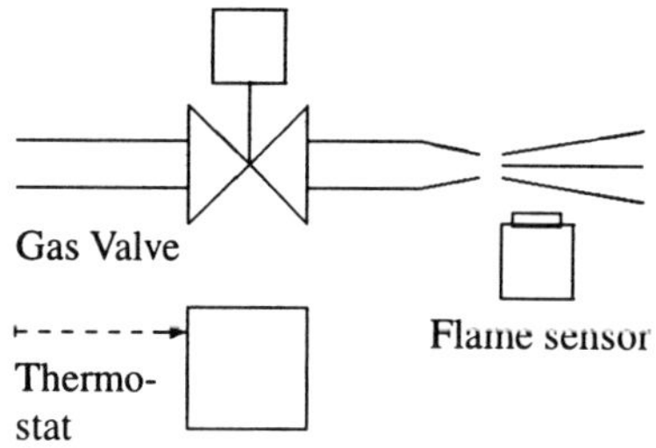

Figure 3. The gasburner

6. The Case Study "Gasburner"

The following case study illustrates how fast and efficient real-time systems can be implemented by the algorithm presented here in comparison with the conventional ProCoS-style. We consider the well-known gasburner case study of the ProCoS-project. The gasburner [12] is triggered by a thermostat; it can directly control a gas valve and monitor the flame (Fig. 3).

This physical system is modelled by three Boolean observables: "hr" (heatrequest) represents the state of the thermostat, "flame" ("flame") represents the presence of a flame at the gas valve, "gas" represents the state of the gas valve. One of the top-level requirements is the formula $\Box(\ell \leq 30 \Longrightarrow \int(\text{gas} \wedge \neg\text{fl}) \leq 4)$ which assures that in every period of 30 seconds gas must not leak for more than four seconds. In the ProCoS-project [3] this case-study was transformed first from Duration Calculus into Implementables. This yielded a specification of a controller using four states idle (id), purge (pg), ignite (ig), and burn (bn) fulfilling the following assertions with an $\varepsilon' > 0$:

$$\lceil\rceil \vee \lceil \text{id} \rceil ; \text{true}$$

$$\lceil \text{id} \rceil \longrightarrow \lceil \text{id} \vee \text{pg} \rceil \qquad \lceil \text{pg} \rceil \longrightarrow \lceil \text{pg} \vee \text{ig} \rceil$$

$$\lceil \text{ig} \rceil \longrightarrow \lceil \text{ig} \vee \text{bn} \rceil \qquad \lceil \text{bn} \rceil \longrightarrow \lceil \text{bn} \vee \text{id} \rceil$$

$$\lceil \neg\text{id} \rceil ; \lceil \text{id} \wedge \neg\text{hr} \rceil \longrightarrow \lceil \text{id} \rceil \qquad \lceil \neg\text{bn} \rceil ; \lceil \text{bn} \wedge \text{hr} \wedge \text{fl} \rceil \longrightarrow \lceil \text{bn} \rceil$$

$$\lceil \neg\text{pg} \rceil ; \lceil \text{pg} \rceil \xrightarrow{\leq 30} \lceil \text{pg} \rceil \qquad \lceil \neg\text{ig} \rceil ; \lceil \text{ig} \rceil \xrightarrow{\leq 1} \lceil \text{ig} \rceil$$

$$\lceil \text{pg} \rceil \xrightarrow{30+\varepsilon'} \lceil \neg\text{pg} \rceil \qquad \lceil \text{ig} \rceil \xrightarrow{1+\varepsilon'} \lceil \neg\text{ig} \rceil$$

$$\lceil \text{id} \wedge \text{hr} \rceil \xrightarrow{\varepsilon'} \lceil \neg\text{id} \rceil \qquad \lceil \text{bn} \wedge \neg\text{hr} \rceil \xrightarrow{\varepsilon'} \lceil \neg\text{bn} \rceil$$

$$\lceil \text{bn} \wedge \neg\text{fl} \rceil \xrightarrow{\varepsilon'} \lceil \neg\text{bn} \rceil$$

The gas valve should be opened iff state is in $\{\text{bn}, \text{ig}\}$.

In the ProCoS-project this specification was transformed via several steps and interfacing languages to hardware. Here we need not perform these transformations. It is sufficient to apply the algorithm. We show the δ tables after steps (1)–(3).

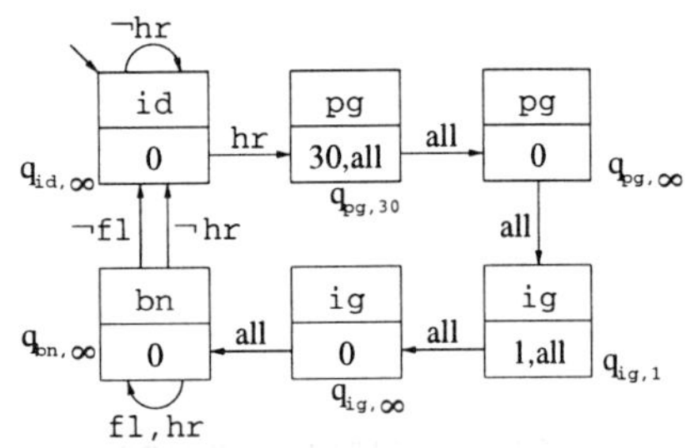

Figure 4. The synthesised implementation of the gasburner.

step (1)	id, ∞	pg, 30	pg, ∞	ig, 1	ig, ∞	bn, ∞
¬hr, ¬fl	id, ~~pg~~	pg, ig	pg, ig	ig, bn	ig, bn	bn, id
¬hr, fl	id, ~~pg~~	pg, ig	pg, ig	ig, bn	ig, bn	bn, id
hr, ¬fl	id, pg	pg, ig	pg, ig	ig, bn	ig, bn	bn, id
hr, fl	id, pg	pg, ig	pg, ig	ig, bn	ig, bn	bn, ~~id~~

step (2)	id, ∞	pg, 30	pg, ∞	ig, 1	ig, ∞	bn, ∞
¬hr, ¬fl	id	pg, ~~ig~~	pg, ig	ig, ~~bn~~	ig, bn	bn, id
¬hr, fl	id	pg, ~~ig~~	pg, ig	ig, ~~bn~~	ig, bn	bn, id
hr, ¬fl	id, pg	pg, ~~ig~~	pg, ig	ig, ~~bn~~	ig, bn	bn, id
hr, fl	id, pg	pg, ~~ig~~	pg, ig	ig, ~~bn~~	ig, bn	bn

step (3)	id, ∞	pg, 30	pg, ∞	ig, 1	ig, ∞	bn, ∞
¬hr, ¬fl	id	pg	~~pg~~, ig	ig	~~ig~~, bn	~~bn~~, id
¬hr, fl	id	pg	~~pg~~, ig	ig	~~ig~~, bn	~~bn~~, id
hr, ¬fl	~~id~~, pg	pg	~~pg~~, ig	ig	~~ig~~, bn	~~bn~~, id
hr, fl	~~id~~, pg	pg	~~pg~~, ig	ig	~~ig~~, bn	bn

The S_e tables are all unchanged and thus omitted here. Figure 4 shows the result of the algorithm applied to the gasburner specification.

7. Future Work

The main advantage of the synthesis algorithm presented is that we gained a new methodology for correct design of real-time systems combining refinement and synthesis. In comparision to this the ProCoS-method would refine from top-level specifications down to Transputers via many intermediate refinement steps which is hardly applicable to realistic problems.

Currently, we are interested in improvements of the synthesis algorithm in order to cope with bigger subsets of DC-formulas than Implementables. Some minor generalisations are worked out already, but we think that it is more important to recognise which kinds of formulas are useful in the UniForM-project than to find slight generalisations of both the algorithm and Implementables.

Acknowledgements

I would like to thank E.-R. Olderog, C. Dietz, S. Kleuker, M. Schenke, and J. Tapken for detailed comments and various discussions on the subject of this paper.

References

[1] R. Alur, C. Courcoubetis, and D. Dill. Model-Checking for Real-Time Systems. In *Fifth Annual IEEE Symp. on Logic in Computer Science*, pages 414–425. IEEE Press, 1990.

[2] R. Alur and D. Dill. A theory of timed automata. *Theoret. Comput. Sci.*, 126:183–235, 1994.

[3] J. Bowen, C. Hoare, H. Langmaack, E.-R. Olderog, and A. Ravn. *ProCoS II: A ProCoS II Project Final Report*, chapter 7, pages 76–99. Number 59 in Bulletin of the EATCS. European Association for Theoretical Computer Science, June 1996.

[4] C. Daws, A. Olivero, S. Tripakis, and S. Yovine. The tool Kronos. In *Hybrid Systems III*, volume 1066 of *LNCS*, pages 208–219. Springer, 1996.

[5] H. Dierks. PLC-Automata: A New Class of Implementable Real-Time Automata. In M. Bertran and T. Rus, editors, *ARTS'97*, volume 1231 of *LNCS*, pages 111–125. Springer, May 1997.

[6] He Jifeng, C. Hoare, M. Fränzle, et al. Provably Correct Systems. In *FTRTFT*, volume 863 of *LNCS*, pages 288–335. Springer, 1994.

[7] C. Heitmeyer and D. Mandrioli, editors. *Formal Methods for Real-Time Computing*. Wiley, 1996.

[8] B. Krieg-Brückner, J. Peleska, E.-R. Olderog, et al. UniForM — Universal Formal Methods Workbench. In *Statusseminar des BMBF Softwaretechnologie*, pages 357–378. BMBF, Berlin, 1996.

[9] C. Lewerentz, editor. *Formal Development of Reactive Systems: Case Study "Production Cell"*, volume 891 of *LNCS*. Springer, 1995.

[10] B. Moszkowski. A Temporal Logic for Multilevel Reasoning about Hardware. *IEEE Computer*, 18(2):10–19, 1985.

[11] A. Ravn. Design of Embedded Real-Time Computing Systems. Technical Report 1995-170, Technical University of Denmark, 1995.

[12] A. Ravn, H. Rischel, and K. Hansen. Specifying and Verifying Requirements of Real-Time Systems. *IEEE Transactions on Software Engineering*, 19:41–55, Jan. 1993.

[13] Zhou Chaochen, C. Hoare, and A. Ravn. A Calculus of Durations. *IPL*, 40/5:269–276, 1991.

A. Semantics of PLC-Automata

The DC-semantics $[\![\mathcal{A}]\!]_{DC}$ of a PLC-Automaton $\mathcal{A} = (Q, \Sigma, \delta, \pi_0, \varepsilon, S_e, S_t, \Omega, \omega)$ is given by the conjunction of the following predicates in the observables $\mathsf{state} : \mathsf{Time} \longrightarrow Q$, $\mathsf{input} : \mathsf{Time} \longrightarrow \Sigma$ and $\mathsf{output} : \mathsf{Time} \longrightarrow \Omega$. First of all, the start of the automaton in the proper initial state is expressed by:

$$\lceil\rceil \quad \vee \quad \lceil \pi_0 \rceil \,;\mathsf{true}. \tag{4}$$

Note that $\lceil \pi_0 \rceil$ is an abbreviation of $\lceil \mathsf{state} = \pi_0 \rceil$. The transition function, the cyclic behaviour, and the output is modelled by the following formulas:[2]

$$\lceil \neg\pi \rceil ; \lceil \pi \wedge A \rceil \longrightarrow \lceil \pi \vee \delta(\pi, A) \rceil \tag{5}$$

$$\lceil \pi \wedge A \rceil \xrightarrow{\varepsilon} \lceil \pi \vee \delta(\pi, A) \rceil \tag{6}$$

$$\Box(\lceil \pi \rceil \Longrightarrow \lceil \omega(\pi) \rceil) \tag{7}$$

where A ranges over sets with $\emptyset \neq A \subseteq \Sigma$. For states without delay requirement ($S_t(\pi) = 0$) we postulate:

$$\lceil \pi \wedge A \rceil \xrightarrow{2\varepsilon} \lceil \delta(\pi, A) \rceil \tag{8}$$

$$\pi \notin \delta(\pi, A) \Longrightarrow \lceil \neg\pi \rceil ; \lceil \pi \wedge A \rceil^{=\varepsilon} \longrightarrow \lceil \neg\pi \rceil \tag{9}$$

For states with delay requirement ($S_t(\pi) > 0$) we have:

$$\lceil \pi \rceil ; \lceil \pi \wedge A \rceil^{=2\varepsilon} \xrightarrow{2\varepsilon + S_t(\pi)} \lceil \delta(\pi, A) \rceil \tag{10}$$

$$\lceil \neg\pi \rceil ; \lceil \pi \wedge A \rceil \xrightarrow{\leq S_t(\pi)} \lceil \pi \vee \delta(\pi, A \setminus S_e(\pi)) \rceil \tag{11}$$

$$\lceil \neg\pi \rceil ; \lceil \pi \rceil ; \lceil \pi \wedge A \rceil^{=\varepsilon} \xrightarrow{\leq S_t(\pi)} \lceil \pi \vee \delta(\pi, A \setminus S_e(\pi)) \rceil \tag{12}$$

$$A \cap S_e(\pi) = \emptyset \Longrightarrow \lceil \pi \wedge A \rceil \xrightarrow{2\varepsilon} \lceil \delta(\pi, A) \rceil \tag{13}$$

$$A \cap S_e(\pi) = \emptyset \Longrightarrow \lceil \neg\pi \rceil ; \lceil \pi \wedge A \rceil^{=\varepsilon} \longrightarrow \lceil \neg\pi \rceil \tag{14}$$

We state two simple properties that are easy to see for a PLC-Automaton $\mathcal{A} = (Q, \Sigma, \delta, q_0, \varepsilon, S_t, S_e, \Omega, \omega)$ with $\pi \in \Omega$ and $\emptyset \neq \varphi \subset \Sigma$:

Lemma A.1 $\lceil \pi \rceil \longrightarrow \lceil \pi \vee \omega(\delta(\omega^{-1}(\pi), \Sigma)) \rceil$.

Lemma A.2 $\lceil \neg\pi \rceil ; \lceil \pi \wedge \varphi \rceil \longrightarrow \lceil \pi \vee \omega(\delta(\omega^{-1}(\pi), \varphi)) \rceil$

B. Correctness of the Algorithm (Thm. 5.5)

Sketch of the Proof: Initialisation: This is obvious by the construction of $\mathcal{A}$ and by (4) and (7).

Sequencing: Suppose $SPEC$ contains a sequencing constraint $\lceil \pi \rceil \longrightarrow \lceil \pi \vee \Gamma \rceil$. We initialised the δ^π-table with Γ_π and no step adds entries, hence by Def. 5.2 and Lemma A.1 we know that this is true.

Unbounded Stability: Suppose $SPEC$ contains an unbounded stability. By step (1) and with Lemma A.2 we can conclude that this also holds.

Bounded Stability: Suppose that $SPEC$ contains a bounded stability $\lceil \neg\pi \rceil ; \lceil \pi \wedge \psi \rceil \xrightarrow{\leq t} \lceil \pi \vee \Psi(\psi) \rceil$. By step (2) we know that only changes to $\Psi(\psi)$ are allowed for the first states $q_{\pi,t_1}, \ldots, q_{\pi,t_k}$ with $t_k = t$ when reading inputs in ψ. By Lemma 5.1 we know that the transition from q_{π,t_i} to $q_{\pi,t_{i+1}}$ can only happen after $S_t(q_{\pi,t_i})$ seconds. By Def. 5.2 we know that $\sum_{i=1}^{k} S_t(q_{\pi,t_i}) = t$ holds.

Synchronisation: Suppose $SPEC$ contains a synchronisation $\lceil \pi \wedge \varphi \rceil \xrightarrow{s} \lceil \neg\pi \rceil$. The negation of this requirement is: $\mathsf{true}; \lceil \pi \wedge \varphi \rceil^{=s} ; \lceil \pi \rceil ; \mathsf{true}$. By finite variability and (7) we can find a finite sequence $q_{\pi,t_j}, \ldots, q_{\pi,t_k}$ with $1 \leq j \leq k \leq n(\pi) + 1$ such that

$$\mathsf{true}; (\lceil q_{\pi,t_j} \rceil ; \ldots ; \lceil q_{\pi,t_k} \rceil \wedge \lceil \pi \wedge \varphi \rceil^{=s}); \lceil \pi \rceil ; \mathsf{true} \tag{15}$$

holds. Due to the construction of the PLC-Automaton we know that there is no transition from q_{π,t_i} to q_{π,t_i} supposed that $i < k$. Hence, we can conclude by (10) that all $\lceil q_{\pi,t_i} \rceil$-phases can not have last longer than $t_i - t_{i-1} + 2\varepsilon$ seconds.

In the case $s > t_k$ we can assign to each $\lceil q_{\pi,t_i} \rceil$ phase an upper time bound by $t_i - t_{i-1} + 2\varepsilon$ and so we get:

$$\mathsf{true}; (\lceil q_{\pi,t_j} \rceil^{\leq t_j - t_{j-1} + 2\varepsilon} ; \ldots ; \lceil q_{\pi,t_k} \rceil^{\leq t_k - t_{k-1} + 2\varepsilon} \wedge \lceil \pi \wedge \varphi \rceil^{=s}); \lceil \pi \rceil ; \mathsf{true}$$

Now we can conclude:

$$\begin{aligned} s &\leq \sum_{i=j}^{k} (t_i - t_{i-1} + 2\varepsilon) = t_k - t_{j-1} + 2(k - j + 1)\varepsilon \\ &\leq t_k + (2(k - 1 + 1))\varepsilon \end{aligned}$$

which is a contradiction to the definition of ε and (3).

In the case that $t_k \geq s$ and $j = k$ holds we know by (8) or (13) in conjunction with $S_e^\pi(\varphi, t_k) = \mathsf{false}$ that the $\lceil q_{\pi,t_k} \rceil$ phase could not last longer than 2ε seconds. This leads to: $\mathsf{true}; (\lceil \pi \wedge \varphi \rceil^{=s} \wedge \lceil q_{\pi,t_k} \rceil^{\leq 2\varepsilon}); \lceil \pi \rceil ; \mathsf{true}$. This can be weakened to $s \leq 2\varepsilon$ which is a contradiction to the definition of ε and (2).

In the case that $j < k$ and $t_k \geq s > t_{k-1}$ holds we know by (9) or (14) in conjunction with $S_e^\pi(\varphi, t_k) = \mathsf{false}$ that the $\lceil q_{\pi,t_k} \rceil$ phase could not last longer than ε seconds. This leads to:

$$\mathsf{true}; (\lceil q_{\pi,t_j} \rceil^{\leq t_j - t_{j-1} + 2\varepsilon} ; \ldots ; \lceil q_{\pi,t_{k-1}} \rceil^{\leq t_{k-1} - t_{k-2} + 2\varepsilon} ; \lceil q_{\pi,t_k} \rceil^{\leq \varepsilon} \wedge \lceil \pi \wedge \varphi \rceil^{=s}); \lceil \pi \rceil ; \mathsf{true}$$

This can be weakened to

$$\begin{aligned} s &\leq \varepsilon + \sum_{i=j}^{k-1} (t_i - t_{i-1} + 2\varepsilon) \\ &= \varepsilon + t_{k-1} - t_{j-1} + 2(k - j)\varepsilon \\ &\leq t_{k-1} + (2(k - 1) + 1)\varepsilon \end{aligned}$$

which is a contradiction to the definition of ε and (3).

Suppose now $j < k$ and $s \leq t_{k-1}$. If $j < k - 1$ holds (15) is a contradiction to (5) with $A = \varphi$ because of $\pi \notin \delta^\pi(\varphi, t_{k-1})$. If $j = k - 1$ we know because of the same reason that the $\lceil q_{\pi,t_{k-1}} \rceil$ phase must be shorter than 2ε seconds due to (13). The $\lceil q_{\pi,t_k} \rceil$ phase could not last longer than ε seconds as in the previous case. Hence, we get $s < 3\varepsilon$ which contradicts (2) again. □

[2] In the formulae we use A as an abbreviation for $\mathsf{input} \in A$ resp. $\delta(\pi, A)$ for $\mathsf{state} \in \{\delta(\pi, a) | a \in A\}$.

A Source-level Dynamic Analysis Methodology and Tool for High-level Synthesis

Chih-Tung Chen and Kayhan Küçükçakar

Unified Design System Laboratory
Motorola, Inc.
{chen,kayhan}@adtaz.sps.mot.com

Abstract

This paper presents a novel source-level dynamic analysis methodology and tool for High-Level Synthesis (HLS). It not only for the first time enables HLS to offer source-level design debugging on the 'synthesized' RTL designs, but also allows the designer to analyze their dynamic characteristics, such as resource utilization, power consumption, etc., at the algorithmic (source) level. This technology has been proven in the industry as the critical element for successfully designing a microcontroller with 300+ instructions with Matisse, an interactive HLS system. Additionally, we demonstrate the use of this technology for architectural power optimization.

1. Introduction

After close to two decades of research, commercial High-level Synthesis (HLS) tools have been introduced recently and have been in some production use. A tutorial on HLS methodology and past research can be found in [7] and [10]. As discussed in [8], the question is not whether high-level synthesis will be successful or not but how widely it will be used in the industry. As pointed out by Bergamaschi in [1], the discussion in [8] and our own study in Motorola, there are some common problems with traditional HLS systems which still hinder designers from embracing HLS. Two specific problems among them are the controllability/predictability and simulation/analysis of the synthesized RTL implementations. First of all, designers demand the controls of the HLS process in order to generate the desired and predictable results. The predictability is found to be crucial for subsequent design tasks including RTL/logic synthesis, data-path module generation, floorplanning, layout, test, and verification. Secondly, designers need an 'intuitive' way to analyze and verify the output of HLS. For example, the cycle-by-cycle behavior of the synthesized design is often different from the one exhibited by the simulation of the input description. Furthermore, the designer needs to understand the internal details, not just the I/O behavior, of the synthesized design so that he/she can change and annotate the subsequent HLS iterations in order to get the desired results.

To accelerate the industrial adoption of HLS, we have developed Matisse, a versatile architectural design tool, which answer many challenges raised in the above studies and discussions [6]. It is beyond the scope of this paper to describe the differences of Matisse from traditional HLS.

This paper presents a novel source-level dynamic execution analysis methodology incorporated in Matisse which enables the designers to analyze the dynamic characteristics of HLS-generated designs in the most intuitive manner: *the original behavioral description.* A portion of this methodology is analogous to the source-level debugging of a compiled software code, where the programmer debugs the execution of the compiled machine code while viewing the original source code. In this analogy, HLS, the input description, and the simulation of the synthesized design correspond to the software compiler, the software source code, and the execution of the compiled machine code, respectively. Additionally, our methodology allows the designers to analyze the design's dynamic characteristics from the source-level execution point of view such as cycle-based resource utilization and power consumption. Thus, the designer can correlate the synthesis results with the HLS input description and constraints, and then, if necessary, adjusts the subsequent HLS iteration to optimize the design's dynamic characteristics more effectively.

Note that our source-level analysis methodology is different from the typical HDL (Hardware Description Language) debugging tools which are built on top of the HDL simulators. These tools allow the designers to debug their HDL code being simulated by the simulator. Applying these tools in the HLS context, the designer is limited to analyze the simulation of the input behavioral description and the RTL implementation **separately** and **independently**. On the contrary, our methodology bridges over this gap by analyzing the RTL simulation from the cycle-accurate execution point of view of the input behavior descrip-

1080-1820/97 $10.00 © 1997 IEEE

tion.

To our best knowledge, the only related work in the hardware design technology is the finite-state machine or flowchart design/analysis tools, such as *VisualHDL* from Summit Design Inc., *DesignBook* from Escalade, *Express* from i-Logix and *speedCHART* from Speed. This type of tools offer visual feedback and controls through state-transition diagrams or flow charts while simulating the HDL code generated for RTL/logic synthesis. The technique for the above FSM/flowchart analysis relies on the fact that there is a clear correlation between the specification and the generated implementation (HDL code); e.g., state registers. However, HLS doesn't offer such straightforward correlation for source-level analysis of HLS designs.

The remainder of this paper is organized as follows. Section 2 gives an overview of what it takes to implement a source-level execution analysis methodology for HLS. In Section 3, we describe the concept of design annotation during HLS and the techniques for enabling the source-level analysis. Section 4 presents the source-level analysis environment in Matisse and illustrates various analysis capabilities offered by this environment. In Section 5, we discuss the applications of our methodology including a microcontroller design and an architectural power optimization experiment. Finally, we conclude this paper in Section 6.

2. Source-level execution analysis methodology

As discussed earlier, the focus of this work is to develop a source-level execution analysis methodology for HLS. The methodology should not only serve as the symbolic debugger for the RTL designs generated by HLS, but also support the analysis of the dynamic characteristics such as cycle-based resource utilization and power consumption. There were several questions raised at the early stage of this work. Is the methodology feasible for HLS? Also, can we leverage existing techniques used for software debugging and FSM/flowchart analysis?

2.1. Feasibility

To understand why it is feasible for HLS to offer the source-level execution analysis, recall that HLS is a process of mapping an algorithmic description of a design behavior into a Register-Transfer Level (RTL) implementation. The HLS tasks include scheduling, resource allocation and sharing, interconnect creation, and finally controller (FSM) generation. Figure 1 shows an example of the mapping performed by HLS. It can be seen that there exists a mapping from the input description to the states or state transitions of the controller (scheduling) and to RTL components and interconnect (binding). Such mapping information can be used to correlate the RTL simulation results back to the input description (the source). Briefly speaking, if we know the current state of the controller (FSM) or the state transition being taken in a particular cycle during RTL simulation, the reverse mapping of the scheduling and binding should tell us at this cycle which portion of the input description is executed and which components and interconnect are utilized. Thus, cycle-based source-level execution and resource utilization become feasible.

2.2. Mathematical model

In this section, we first describe the essence of the HLS design execution, meaning the simulation of the RTL implementation. Then, we define the reverse mapping of the scheduling and derive the source-level execution from the above information. Finally, some extensions of the model will be addressed.

For the sake of simpler explanation, assume the HLS design contains a single FSM as the controller. If the generated controller is *state-based* (e.g., a Moore-type FSM)

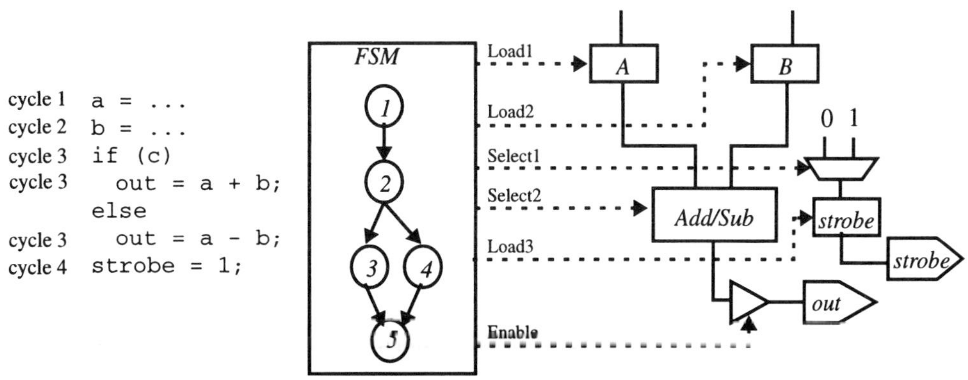

a. Specification *b. RTL Implementation*

Figure 1: High-level synthesis process

with states $S = \{s_1, \ldots, s_A\}$, the execution of the HLS design, therefore, can be characterized by an ordered list of states $DX = \{dx_1, \ldots, dx_B\}$ such that $dx_b \in S$ and (dx_b, dx_{b+1}) is a valid state transition. Note that dx_b represents the current state of the controller at the bth cycle of the HLS design execution.

Let the input description be a list of statements $I = \{i_1, \ldots, i_C\}$. The scheduling task of HLS can be represented by a one-to-many mapping $\lambda: I \rightarrow S^S$ such that $\lambda(i_x)$ is a subset of S and denotes the states in which statement i_x is to be executed. Thus, the reverse mapping of λ becomes $\lambda^{-1}: S \rightarrow I^I$ where $\lambda^{-1}(s_y) = \{i_x \mid s_y \in \lambda(i_x)\}$.

Formally, the cycle-based source-level execution can be represented by an ordered list $SX = \{sx_1, \ldots, sx_D\}$ such that sx_d is a subset of statements S which executed at the dth cycle of the HLS design execution. It can be seen easily now that the source-level execution SX can be derived from DX and λ^{-1} as $\{\lambda^{-1}(dx_1), \ldots, \lambda^{-1}(dx_D)\}$ since sx_d is equal to $\lambda^{-1}(dx_d)$.

On the contrary, if the generated controller is state-transition-based (e.g., a Mealy-type FSM) with state transitions $T = \{t_1, \ldots, t_E\}$, the execution of the HLS design must be characterized by an ordered list of state transitions $DT = \{dt_1, \ldots, dt_F\}$ such that $dt_f \in T$ and (dt_f, dt_{f+1}) is connected through a valid state. The HLS scheduling, in this situation, is a one-to-many mapping $\lambda: I \rightarrow T^T$ such that $\lambda(i_x)$ is a subset of T in which statement i_x is to be executed. Additionally, the reverse mapping λ^{-1} becomes $\lambda^{-1}: T \rightarrow I^I$ where $\lambda^{-1}(t_y) = \{i_x \mid t_y \in \lambda(i_x)\}$. Similarly, the source-level execution SX can be derived from DT and λ^{-1} as $\{\lambda^{-1}(dt_1), \ldots, \lambda^{-1}(dt_D)\}$ since sx_d is equal to $\lambda^{-1}(dt_d)$.

The model above can be easily extended to support the cycle-based resource utilization. Briefly speaking, the cycle-based resource utilization is a dynamic analysis of data-path resource utilization throughout the run (simulation) time of the HLS design. The data-path resource can be a data-path component or an interconnect. The advantage of analyzing cycle-based resource utilization is that it can accurately portray the actual resource utilization of designs with complex conditional resource sharing and data-dependent loops.

Like source-level execution, the cycle-based resource utilization can be derived from the HLS design execution and the reverse mapping of the HLS binding. For example, assume the HLS design is state-based as described earlier and comprises of a set of resources $R = \{r_1, \ldots, r_G\}$. The HLS binding can be transformed to a one-to-many mapping $\gamma: R \rightarrow S^S$ such that $\gamma(r_x)$ is a subset of S in which resource r_x is to be used, and its reverse mapping becomes $\gamma^{-1}: S \rightarrow R^R$ where $\gamma^{-1}(s_y) = \{r_x \mid s_y \in \gamma(r_x)\}$. Finally, let the cycle-based resource utilization be an ordered list $U = \{u_1, \ldots, u_H\}$ such that u_h is a subset of R which are utilized at the hth cycle of the HLS design execution. Thus, U can be derived as $\{\gamma^{-1}(dx_1), \ldots, \gamma^{-1}(dx_H)\}$ since u_h is equal to $\gamma^{-1}(dx_h)$.

Although the model described above assumes a single-FSM controller, it can also be extended to support those HLS systems which generate multi-FSM controllers such as Hercules/Hebe [5] and Matisse [6]. For example, assume the FSMs are all state-based and use the same clock. Let the FSMs be $M = \{m_1, \ldots, m_I\}$. The HLS design execution can be characterized by using multiple DX lists, one for each FSM as described earlier. The source-level execution SX can then be derived from λ^{-1} and the DX lists with sx_d as $\lambda^{-1}(dx_d^1) \cup \ldots \cup \lambda^{-1}(dx_d^I)$.

Another extension of our model is to use operations and values instead of statements for the source-level analysis. This extension would allow the user to better analyze intra-statement events. This can be accomplished simply by replacing the statement list I with a list containing individual operations and values.

2.3. Analysis flow

As described in Section 2.2, our source-level execution analysis methodology is enabled by two key ingredients: the reverse mapping and the state (or state-transition) history of the underlying controller which captures the essence of the HLS design execution. In this section, we will discuss what it takes to incorporate such a methodology into an HLS system and the associated analysis flow.

Theoretically, an HLS design can be simulated "as is" and its state-based execution history still can be obtained through the simulator by monitoring the state registers of the controller. However, the transition-based execution history cannot be captured by such monitoring. For example, assuming there are two states s_1 and s_2 with two conditional transitions from s_1 to s_2 depending on the value of an input *cond*. Knowing that the HLS design is in state s_2 is not sufficient to tell which transition the design took to reach s_2. The problem is further complicated by the potential logic glitches, such as those on *cond*, which may cause specious state transitions in the middle of a cycle.

Therefore, we introduce the concept of *design annotation* and *analyzable HLS designs*. The design annotation is a process of embedding one or more daemons into the design during HLS. The embedded daemons when simulated will use specific techniques to collect dynamic data of interest such as state transitions. Moreover, the embedded daemons will not affect the functional behavior of the design during simulation. The HLS design with the embedded daemons is called *source-level analyzable*. The design annotation will be further discussed in Section 3.

The second crucial task in our methodology is to generate the reverse mapping of scheduling and binding. This involves reprocessing the scheduling and binding decisions and storing the information tabularly with states or

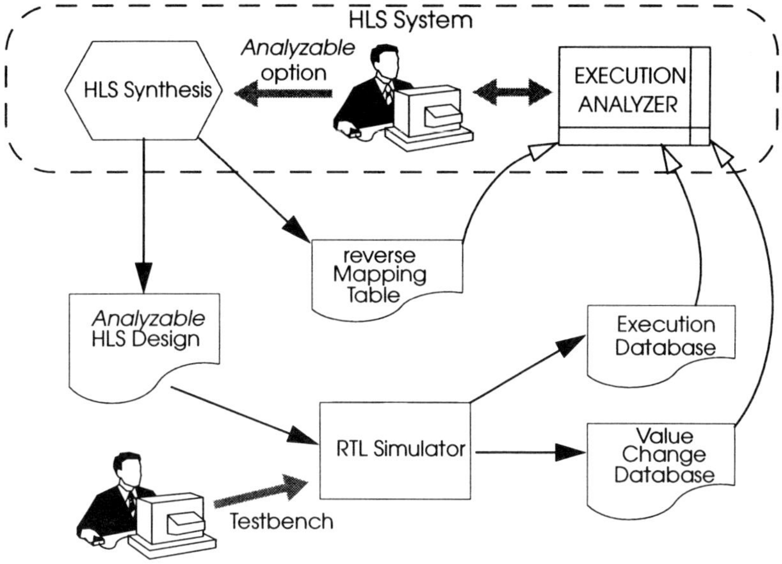

Figure 2: Source-level execution analysis flow

state transitions as the indices. This task, however, is best carried out during the HLS controller generation where the complete scheduling and binding information is available and the states as well as the state-transitions are being generated.

As shown in Figure 2, a typical source-level execution analysis flow begins with synthesizing an *analyzable* HLS design containing embedded daemons and generating one or more tables containing the reverse mapping of scheduling and binding.

Following that, the analyzable HLS design is simulated, and an execution database is generated by the embedded daemons along with a value change database.

The execution database stores the execution history; e.g. a list of 2-tuple (*t*, *st*) where *t* is the simulation time and *st* is an identifier of a particular state transition. The value change database contains the value history of one or more signals in the design. This value change database enables our methodology to offer both behavioral and structural signal analysis along with the source-level execution analysis. Although the generation of the above databases is needed for post-simulation analysis and time-backward analysis, the real-time source-level analysis is also possible by using the simulator's application programming interface (API) to generate and access the records in real time.

The last step in the analysis flow is to invoke a source-level execution analysis tool similar to a software symbolic debugger for the actual analysis work. The details of such source-level execution analysis tool will be described in Section 4.

3. Design annotation

Design annotation is a process of embedding one or more daemons into the design during HLS for collecting dynamic data of interest using specific techniques. Currently, we identified two types of design annotation: controller annotation for capturing the state-transition based execution history and data-path annotation for calculating cycle-based power consumption.

Note that the design annotation should be performed such that the analyzable HLS design doesn't incur any overhead in the downstream synthesis tools as compared to the original HLS design. In our case, we hide the embedded daemons from logic synthesis by enclosing them with directives that turn logic synthesis off and on.

3.1. Controller annotation

As discussed in Section 2.3, the transition-based execution history cannot be captured by simply recording the controller's state registers. With potential logic glitches, it is also difficult to derive the actual transition history by continuously monitoring the state-transition logic of the controller. Since *Matisse*, our targeted HLS system, is state-transition based, we annotated the controller with one daemon per FSM. Each daemon comprises of a state-transition identification (*STI*) logic and an *STI* register, and is connected with a typical Mealy-type FSM as shown in Figure 3. Like other Mealy-type output functions, the *STI* logic is a function of FSM's current state *CS* and the input *I*, and computes the next state-transition identification *NSTI*. The *STI* register is analogous to the FSM state register. Functionally speaking, the *STI* register represents edge-triggered flip-flops driven by the FSM clock signal.

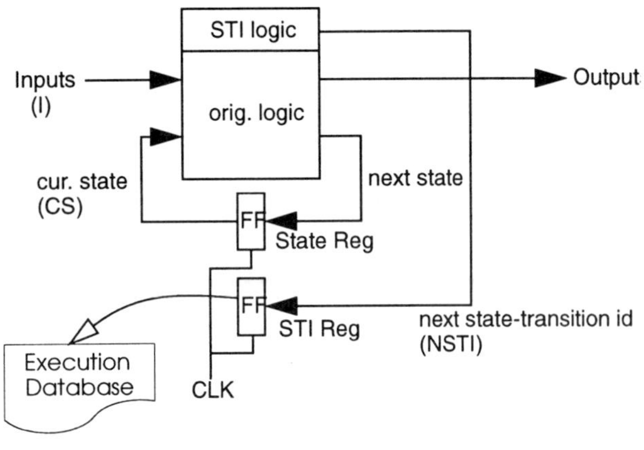

Figure 3: STI annotation

Also shown in Figure 3 is an arc connecting the *STI* register output to the execution database. It points out the real role of the *STI* register, which is to store the identifier of the actual state transition committed by the FSM into the execution database along with the simulation time at the end of every clock cycle. Thus, the cycle-based state-transition history can be faithfully recorded under the presence of logic glitches.

We found that it is more efficient to generate the reverse mapping tables of scheduling and binding during the controller annotation. Recall that these reverse mapping tables for the state-transition based HLS designs correlate each state transition with a set of input statements being executed or a set of resource being utilized. By using the state transition identifiers generated during the controller annotation as the key field for the reverse mapping tables, the reverse mapping for source-level execution and cycle-based utilization can be done in linear time.

3.2. Data-path annotation

The data-path annotation for cycle-based power analysis is a process to embed a daemon (*SD*) to an individual data-path component or net for gathering signal switching statistics periodically. As shown in Figure 4, *SD* is driven

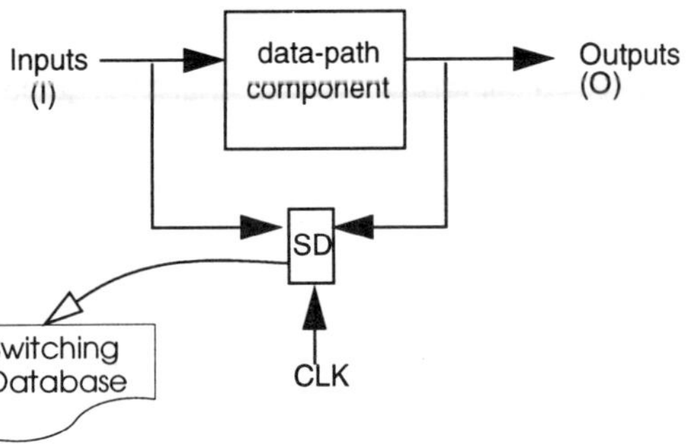

Figure 4: Data-path annotation

by a system or component-specific clock *CLK*. Similar to the *STI* daemon, *SD* stores the peripheral switching statistics of the component into the switching database along with the simulation time at the end of every cycle. Thus, cycle-based power consumption can be computed by the switching statistics in each cycle and the power characterization models of individual components available from RTL power analysis tools such as [11] and [9].

4. Source-level analysis environment

Figure 5 shows the source-level analysis environment developed in Matisse using the methodology described earlier. Currently, it offers source-level execution analysis, dynamic resource utilization analysis and dynamic profiling. We would like to remind that although our source-level analysis environment seemingly resembles a typical HDL debugging tool, the underlying techniques are fundamentally different. More importantly, traditional HDL debugging tools cannot be used for source-level analysis of synthesized designs.

4.1. Source-level execution analysis

As shown in Figure 5, the source-level execution analyzer (*EA*) consists of a control panel, a source window, and a data display window. The control panel allows the designer to perform typical debugging operations such as *start, stop at, next, continue* and *data display.* Due to the fact that the execution history is available in the execution database, *EA* also offers backward stepping (*prev*) and backward continuation to move the simulation time back to a prior clock cycle. The source window is for listing the source code of the HLS design (currently an algorithmic subset of Verilog HDL) and highlighting those statement being executed at the current cycle. It also gives the designer the access to individual statements for setting the stop points and the access to variables and operations for displaying values. Finally, the data display window contains a list of data display entries which are updated automatically whenever the simulation time changes.

The basic core of *EA* can be described in pseudo code as following.

```
import(execution_database, reverse_mapping_tables);
while (not quit) begin
    request_or_compute(next_cycle);
    query_execution_database(next_cycle, assoc_stis);
    exec_stmts = ∅; used_rsrcs = ∅;
    foreach sti in assoc_stis begin
        exec_stmts |= schedule_reverse_mapping(sti);
        used_rsrcs |= binding_reverse_mapping(sti);
    end
    display_time(next_cycle);
    source_window_highlight(exec_stmts);
    structure_window_highlight(used_rsrcs);
end
```

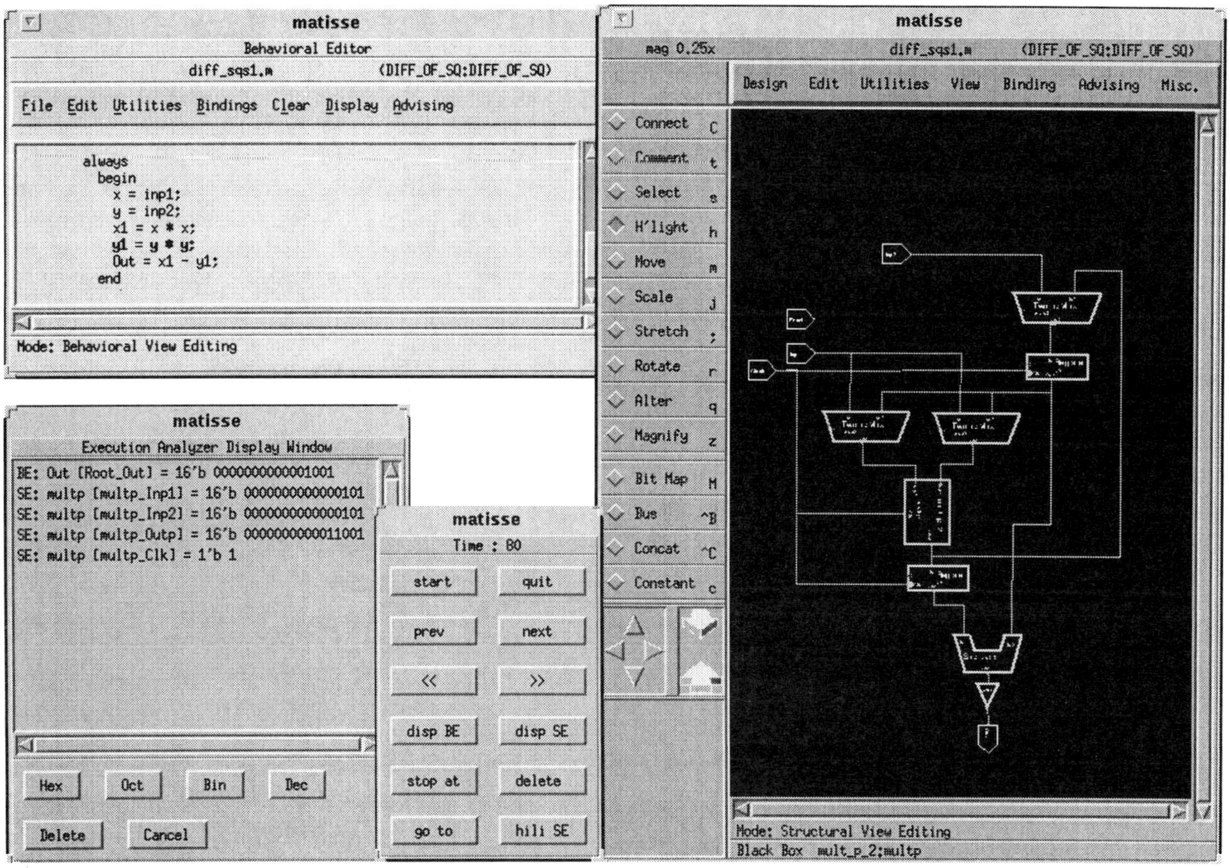

Figure 5: Source-level analysis environment

Note that *sti* stands for state-transition identifier and '|=' is a union-assign operation.

4.2. Dynamic resource utilization analysis

Also shown in Figure 5 is a structure window for cycle-based resource utilization analysis. The structure window of *EA* displays the data-path schematics and highlights the active (utilized) resources at the current cycle in a similar fashion to *SEE-SAW* [2]. The designer can also select a particular net or component through the structure window for value display.

By using the dynamic resource utilization analysis, the designer can visualize 'where' and 'when' a portion of the data path is idle under the actual operational environment. Then the schedule and/or bindings can be adjusted to improve the overall dynamic utilization, or power saving techniques such as signal gating can be applied to reduce the power consumption.

4.3. Cycle-based power analysis

The cycle-based power analysis can be performed by integrating the component-level power models provided by [9] with the data-path annotation technique described in Section 3.2. The intuitive functionality is to provide the designer so-called *power display* similar to the existing *data display* with the same granularity such as per net, per component or even per operation basis. The power display entries are to be shown in the display window and updated whenever the simulation time change.

4.4. Dynamic profiling analysis

Another related analysis available is the dynamic profiling analysis. The dynamic profiling analysis reports the execution frequency of individual statements in the input description and the usage frequency of individual resources in the data path under a scenario of the HLS design simulation. The profiling data is computed in a similar way as described in Section 2.2 for cycle-based source-level analysis using the execution database and the reverse mapping tables except that the computation is accumulated over the entire execution history.

5. Applications of source-level analysis

Matisse equipped with the source-level execution analysis presented in this paper has been used in several industrial designs for evaluation or production. The designs included hardwired DSP, encryption/decryption engine, bus interface, control-dominated peripheral, and a microcontroller. The source-level execution analysis was proven to be an invaluable tool to the designers especially for the designs involving considerably interactive and/or incremental design. In this section, we will summarize two notable applications of the source-level analysis.

5.1. Design of an industrial microcontroller

Matisse has been used to redesign a commercial 8-bit microcontroller with 300+ instructions. During this redesign project, it became inevitable that the designers had to schedule the majority of the micro-instructions and mapping them onto a well-defined data-path architecture interactively in order to ensure the compatibility and the efficiency. Thus, they devised an incremental design flow comprising of adding new instructions, interactive micro-instruction scheduling, mapping onto the existing data path, generating the RTL implementation, simulating and debugging. Without using the source-level execution analyzer presented in this paper, the designers will be forced to verify the newly added instructions by analyzing the RTL simulation directly, which was shown to be an overwhelmingly time-consuming task. The designers estimated that our source-level execution analyzer provided 3-4X reduction in debugging time than traditional RTL analysis.

5.2. Architectural power optimization

We performed an experiment of a G.721 ADPCM predictor [3]. We explored low-power architecture in an iterative design flow including architectural design using Matisse, hot point analysis using an RTL power estimator [9], cycle-based resource utilization analysis and profiling as presented in this paper. By identifying the hot points and their utilization patterns, we were able to obtained 4.4X power saving in one person week through power reduction techniques like signal gating, pipelined multipliers, register files, and pipelined scheduling. The results were then verified by an accurate gate-level power estimator [4].

6. Conclusion

As mentioned in Section 1, HLS needs to improve the controllability/predictability and simulation/analysis of the synthesized RTL implementations in order to increase its adoption in the industry. We have presented a unique source-level analysis methodology and tool for HLS to address these problems. It not only for the first time enables HLS to offer an intuitive design debugging capability, but also allows the designer to perform dynamic analysis of resource utilization, power consumption, etc. at the source level. This technology has been shown to be the critical element for effectively applying the HLS methodology on the industrial designs.

7. References

[1] R.A. Bergamaschi, "Productivity Issues in High-level Design: Are Tools Solving the Real Problems?", *In Proceedings of the Design Automation Conference,* pages 674-677, June 1995.

[2] R. Blackburn, D.E. Thomas and P. Koenig, "CORAL II: Linking Behavior and Structure in an IC Design System", *In Proceedings of Design Automation Conference*, pages 529-535, June 1988.

[3] C.-T. Chen and K. Kucukcakar, "An Architectural Power Optimization Case Study using High-level Synthesis", To appear in *Proceedings of International Conference on Computer Design*, October, 1997.

[4] B. George, G. Yeap, M.G. Wloka, S.C. Tyler and D. Gossain, "Power Analysis for Semi-Custom Design", *In Proceedings of Custom Integrated Circuits Conference*, May 1994, pp. 249--252.

[5] D. Ku and G. De Micheli, *High-Level Synthesis of ASICs under Timing and Synchronization Constraints*, Kluwer Academic Publishers, 1992.

[6] *Matisse User Manual,* Motorola, Inc., 1995.

[7] M.C. McFarland, A.C. Parker and R. Camposano, "The High-Level Synthesis of Digital Systems", *In Proceedings of the IEEE*, 78(2):301-318, February 1990.

[8] Panel: Real Requirements of High-level Synthesis, *International Symposium on System Synthesis*, 1996.

[9] *Taos User Manual*, Motorola, Inc.,1996.

[10] R.A. Walker and R. Camposano, *A Survey of High-Level Synthesis Systems,* Kluwer Academic Publishers, 1991.

[11] *WattWatcher/Architect,* Sente, Inc., 1996.

Author Index

IEEE COMPUTER SOCIETY

IEEE Computer Society Publications

The world-renowned Computer Society publishes, promotes, and distributes a wide variety of authoritative computer science and engineering texts. These books are available in two formats: 100 percent original material by authors preeminent in their field who focus on relevant topics and cutting-edge research, and reprint collections consisting of carefully selected groups of previously published papers with accompanying original introductory and explanatory text.

Submission of proposals: For guidelines and information on Computer Society books, send e-mail to cs.books@computer.org or write to the Acquisitions Editor, IEEE Computer Society, P.O. Box 3014, 10662 Los Vaqueros Circle, Los Alamitos, CA 90720-1314. Telephone +1 714-821-8380. FAX +1 714-761-1784.

IEEE Computer Society Proceedings

The Computer Society also produces and actively promotes the proceedings of more than 130 acclaimed international conferences each year in multimedia formats that include hard and softcover books, CD-ROMs, videos, and on-line publications.

For information on Computer Society proceedings, send e-mail to cs.books@computer.org or write to Proceedings, IEEE Computer Society, P.O. Box 3014, 10662 Los Vaqueros Circle, Los Alamitos, CA 90720-1314. Telephone +1 714-821-8380. FAX +1 714-761-1784.

Additional information regarding the Computer Society, conferences and proceedings, CD-ROMs, videos, and books can also be accessed from our web site at http://computer.org/cspress

4/15/97